# LEADING HOLISTICALLY

*Leading Holistically* explores systems thinking in educational leadership—a comprehensive framework that enables leaders to improve their practice by taking a holistic perspective, instead of relying on a one-size-fits-all solution to discrete aspects of their organization. Aiming to expand the existing literature on systems thinking in educational leadership and policy, renowned educational leadership scholars come together in this valuable book to examine systems thinking at the school, district, and state/national levels, providing strategies to guide educators toward success. This important book unpacks the complexity and nuances of systems thinking in educational leadership and policy, helping educators face the growing complexity, change, and diversity in education to realize the promise of improvement for all those connected to and involved in the important endeavor of education.

**Haim Shaked** is Vice President for Academic Affairs and Head of the Department of Education at Hemdat Hadarom College of Education, Netivot, Israel.

**Chen Schechter** is Editor-in-Chief of the *Journal of Educational Administration* and Professor of Leadership, Organizational Development, and Policy in the School of Education at Bar Ilan University, Ramat-Gan, Israel.

**Alan J. Daly** is Chair and Professor of Education Studies at the University of California, San Diego, USA.

# LEADING HOLISTICALLY

## How Schools, Districts, and States Improve Systemically

*Edited by*
*Haim Shaked, Chen Schechter and Alan J. Daly*

Routledge
Taylor & Francis Group
NEW YORK AND LONDON

First published 2019
by Routledge
711 Third Avenue, New York, NY 10017

and by Routledge
2 Park Square, Milton Park, Abingdon, Oxon, OX14 4RN

*Routledge is an imprint of the Taylor & Francis Group, an informa business*



*Library of Congress Cataloging-in-Publication Data*
A catalog record for this title has been requested

ISBN: 978-1-138-55991-2 (hbk)
ISBN: 978-1-138-55992-9 (pbk)
ISBN: 978-0-203-71218-4 (ebk)

Typeset in Bembo
by Swales & Willis Ltd, Exeter, Devon, UK

# CONTENTS

# EDITORS

**Haim Shaked**

Haim Shaked, Ph.D., is Vice President for Academic Affairs and Head of the Department of Education at Hemdat Hadarom College of Education, Netivot, Israel. As a scholar-practitioner with seventeen years of experience as school principal, his research interests include instructional leadership, system thinking in school leadership, and education reform. His book (co-authored by Chen Schechter) *Systems Thinking for School Leaders: Holistic Leadership for Excellence in Education* was published recently by Springer. haim.shaked@hemdat.ac.il

**Chen Schechter**

Chen Schechter, Ph.D., is a Professor of Leadership, Organizational Development and Policy in Education, in the School of Education at Bar-Ilan University, Israel. Chen is the Editor-in-Chief (co) of the *Journal of Educational Administration*. He serves as Chair of the Board of Directors at the Institute for the Advancement of Teaching, Learning and Social Integration. He is also the Chair of AERA Organizational Theory SIG. Chen's research interests include leadership development/preparation, educational change/reform, learning from success, and systems thinking leadership. Chen.Schechter@biu.ac.il

**Alan J. Daly**

Alan J. Daly, Ph.D. is a Professor in the Department of Education Studies at the University of California, San Diego, who specializes in social network analysis and educational change processes. He has authored several books on the topic including one published by Harvard Press, entitled *Social Network Theory and Educational Change*. Professor Daly is also a Fulbright Global Scholar, having spent time in New Zealand and South Africa. ajdaly@ucsd.edu

# CONTRIBUTORS

**Janice Bradley**

Janice Bradley is an Assistant Director for the Utah Educational Policy Center (UEPC) at the University of Utah. Her UEPC work focuses on facilitating school change and improvement, designing professional learning environments and experiences for schools, administrators, and coaches, developing effective collaborative teams and professional learning communities, and building capacity for supportive communication climates. She also serves as an Adjunct Assistant Professor, teaching courses in the Department of Educational Leadership & Policy. janice.bradley@utah.edu

**Pascale Benoliel**

Pascale Benoliel, Ph.D. is a Lecturer and faculty member in the Department of Educational Leadership and Policy, School of Education, Bar-Ilan University, Israel. Her research areas include school leadership, teamwork, team leadership, cross-cultural analysis, comparative research, and quantitative research methods. She has published articles in a range of highly ranked scholarly journals in the fields of psychology and management, as well as in the educational field. pascale.benoliel@biu.ac.il

**Alex J. Bowers**

Alex J. Bowers is an Associate Professor of Education Leadership at Teachers College, Columbia University. His research interests include organizational level data analytics, organizational behavior, school and district leadership, data-driven decision making, high school dropouts and completion, educational assessment and accountability, education technology and school facilities financing. bowers@tc.columbia.edu

**Ebony N. Bridwell-Mitchell**

Ebony N. Bridwell-Mitchell is an Associate Professor at the Harvard Graduate School of Education. She holds a Ph.D. in Organizational Theory and Management from the New York University Stern School of Business and a Master's degree in Public Policy from the Harvard John F. Kennedy School of Government. She leverages her training in organizational management and education policy to study the micro-social and cognitive dynamics of education policy implementation in U.S. public schools. ebony_bridwell-mitchell@gse.harvard.edu

**James Coviello**

James Coviello is an Assistant Professor of Educational Leadership at Saint Joseph's University and a 2017 Graduate Fellow in Public Policy at the University of Texas' Archer Center in Washington, D.C. His research focuses on the interaction of leadership, equity, and policy implementation. Before pursuing his doctorate, James was a high school history teacher for ten years. james.coviello@sju.edu

**Miguel del Fresno**

Miguel del Fresno, Ph.D. is an Associate Professor at the Universidad Nacional de Educacion a Distancia (Spain) and teaches in the Masters of Communication program at Universidad de Navarra (UN), UPV (Bilbao), UCM (Madrid), and University of Sevilla in Spain, as well as a Visiting Professor at University of the Republic of Uruguay (Montevideo) and Visiting Researcher at the U.C. Berkley and U.C. San Diego. mdelfresno@der.uned.es

**David E. DeMatthews**

David E. DeMatthews is an Associate Professor in the Department of Educational Leadership and Policy at the University of Texas (UT) at Austin. David serves as the Program Head of the UT-Austin Urban Principal Leadership Academy and holds a courtesy appointment in the Department of Special Education. Prior to working in higher education, David has worked with urban districts as a high school teacher, middle school administrator, and district administrator. ddematthews@austin.utexas.edu

**Christina J. Dixon**

Christina J. Dixon has devoted her career to supporting education, healthcare, and non-profit leaders in using continuous improvement principles to transform their systems to better to meet the needs of the people they serve. She currently works with the Carnegie Foundation for the Advancement of Teaching as an Associate for Networked Improvement Science, and is an Ed. D. candidate at the University of Virginia. cd3dr@virginia.edu

**Lawrie Drysdale**

Lawrie Drysdale is an Associate Professor in educational leadership within the Melbourne Graduate School of Education at the University of Melbourne, and has a 40 year background in teaching, human resource development, lecturing and research. He coordinates postgraduate courses in educational management, and teaches subjects in leadership, human resource management, marketing, school effectiveness and improvement. He is a member of the International Successful School Principalship Project (ISSPP). drysdale@unimelb.edu.au

**David Eddy-Spicer**

David Eddy-Spicer, Associate Professor at the Curry School of Education, University of Virginia, studies the intersection of accountability policies and the development of interorganizational capacity for continuous improvement in schools and school systems. He is co-editor of the SAGE Handbook of School Organization and has published a range of peer-reviewed journal articles, book chapters, and text and multimedia case studies on professional learning and organizational change in the public sector. dhe5f@virginia.edu

**Jennifer Esswein**

Jennifer Esswein is a Senior Researcher at Education Northwest in the Center for Research, Evaluation and Analysis. She works closely with schools, districts, and state departments of education to increase capacity to use data for making informed policy decisions. Most recently, she has supported states in planning and implementation of new federal policy, specifically with identification of the lowest performing schools in need of support and improvement. Jennifer.Esswein@educationnorthwest.org

**Cori Groth**

Cori Groth is the Associate Director for the Utah Educational Policy Center (UPEC) at the University of Utah, where she is involved in educational research and school improvement at both the preK-12 and higher education levels. She is also an Adjunct Professor with the Department of Educational Leadership and Policy. She earned a Ph.D. and B.S. from the University of Utah and a M.A. from Arizona State University. cori.groth@utah.edu

**David Gurr**

David Gurr is an Associate Professor in educational leadership within the Melbourne Graduate School of Education at the University of Melbourne, and has a 38-year background in secondary teaching, educational psychology, school supervision, and research in educational leadership. He has written extensively on successful school leadership, and is a member of the leadership group of the International Successful School Principalship Project. d.gurr@unimelb.edu.au

**Lisa A.W. Kensler**

Lisa A. W. Kensler is the Emily R. and Gerald S. Leischuck Endowed Professor and Program Coordinator for Educational Leadership in the College of Education at Auburn University. Her original training in ecology continues to fuel her love of systems thinking. Her current research focuses on green schools and the leadership and learning required for transforming schools into more socially just, ecologically responsible, and economically viable communities. lak0008@auburn.edu

**Sharon D. Kruse**

Sharon D. Kruse is Academic Director and Professor at Washington State University, Vancouver. Her scholarship addresses two concerns helping educators better understand the key role leadership plays in schools and exploring how education is currently structured and influenced by social and organizational complexity. By focusing on the ways issues are framed, decisions are made, and problems are identified, she seeks to understand how schools can better meet the needs of students. sharon.kruse@wsu.edu

**Rachael B. Lawrence**

Rachael B. Lawrence is a Postdoctoral Research Fellow at the Center for Education Policy at the University of Massachusetts Amherst. As a program evaluator and an education researcher, she explores school change initiatives; professional learning and development; and, linkages between evaluation and policy decision. She currently serves as Managing Editor of the American Journal of Evaluation. Her Ph.D. in Education Policy & Leadership is from the University of Massachusetts Amherst. rblawren@umass.edu

**Nechama Nadav**

Nechama Nadav is a doctoral student in Educational Leadership, Administration and Policy at School of Education, Bar-Ilan University. She has practical experience in education for more than 20 years that she combines in her research. Her research field deals with system thinking in school leadership and her current research focuses on understanding the characteristics of systems thinking and how systems thinking influences middle leaders. nadav9206@gmail.com

**Xinyu Ni**

Xinyu Ni is a doctoral student in Statistics, Measurement and Evaluation at Teachers College, Columbia University. She is interested in educational data mining, school and district leadership applied psychometric techniques, and educational assessment and evaluation. xn2115@tc.columbia.edu

**Adam Nir**

Adam Nir is a Professor of Educational Administration Policy and leadership and the Abraham Shiffman Chair in Secondary Education at the Hebrew University of Jerusalem Israel. He is the former Chair of the Department of Education at the Hebrew University and the former elected President of the International Society for Educational Planning (ISEP). His research interests include school autonomy, School-Based Management, educational planning, leadership and human resource management in public education. adam.nir@mail.huji.ac.il

**Vicki Park**

Vicki Park is an Assistant Professor of Educational Leadership at San Diego State University. The guiding goal of her work has been to bridge the research-to-practice and practice-to-research gap in order to improve educational opportunities and outcomes for diverse student populations. Her scholarship broadly focuses on assets-based leadership practices for equity, professional collaboration, and organizational learning for continuous improvement. Prior to earning her doctorate, she worked as an elementary and middle school teacher. vpark@sdsu.edu

**Sharon F. Rallis**

Sharon Rallis, Dwight W. Allen Distinguished Professor of Education Policy and Reform at the University of Massachusetts-Amherst, teaches inquiry, program evaluation, qualitative methods, and organizational theory. Her research and program evaluation work has taken her throughout the United States and the world. Her eleven published books include Learning in the Field: An Introduction to Qualitative Research (Sage) and The Research Journey: Introduction to Inquiry (Guilford). She earned her doctorate from Harvard University. sharont@educ.umass.edu

**Andrea Rorrer**

Andrea Rorrer is a Professor in the Department of Educational Leadership and Policy Studies, Director of the Utah Education Policy Center at the University of Utah, and Associate Dean for Research in the College of Education at the University of Utah. Andrea's expertise and scholarship is school district and state leadership, policy, and educational change, particularly those changes aimed at increasing equity in student access and outcomes. Andrea.Rorrer@utah.edu

**Jonathan A. Supovitz**

Jonathan A. Supovitz, Ed.D. is a Professor of Education Policy and Leadership at the University of Pennsylvania's Graduate School of Education and the Director of the Consortium for Policy Research in Education (CPRE). He is also the

Executive Director of the CPRE Knowledge Hub (cprehub.org), a virtual space that brings together researchers, policymakers, and practitioners learn and collaborate on challenges in the education field. jons@gse.upenn.edu

**Cynthia L. Uline**

Cynthia L. Uline is Professor Emeritus of Educational Leadership at San Diego State University. Cynthia serves as Director of SDSU's National Center for the 21st Century Schoolhouse. Her research explores the influence of built learning environments on students' learning, and the roles leaders, teachers, and the public play in shaping learning spaces. Her current research considers the potential of green schools as student-centered, ecologically responsive, and economically viable places for learning. culine@sdsu.edu

# FOREWORD

## Systems Theory: The Friend that Never Arrives

*Michael Fullan*

Systems theory had been formally around for almost three quarters of a century when Kenneth Boulding (1956) wrote *General Systems Theory*. It was supposed to help us understand the dynamics of complex systems, especially global societies and their interactions. Then decade after decade society got ever more complex. You would think that systems theory would thrive as societies offered up ever more dynamic realities. A few did respond. Peter Senge and his colleagues are devoting their entire careers to mapping out systems theory and its use. We do have some books on "understanding systems theory," "system theory tool kits" and the like. And we do have a whole movement of scholars who continue to examine and analyze chaos theory, butterfly effects and other strange attractors.

Overall, however, I would argue that systems theory has not grown commensurate with the problem; and especially on the applied side it has had little growth and "sticking power." It seems that the more complex things get, the more simple the solutions that politicians propose (and evidently the public like). Since society is rife with complexity, and since the problems and consequences are ever more threatening and incomprehensible, we need a movement that begins to tackle applied problems at all levels with complexity thinking and action.

It is for all these reasons that I welcome *Leading Holistically*. There is simply no other book like it. It is reasonably brief, but it covers "the whole system," so to speak. There are three sections of the book and each one has four chapters. The three parts in total examine the entire education system: four chapters on systems thinking at the school level; four on systems thinking at the district/regional level, and four at the state or national level. All twelve chapters are steeped in applied issues, and systems thinking to help us understand what is happening and to advance the cause of systems understanding and improvement.

Right away in Chapter 1 the reader is confronted with the challenge of "why state reforms fail to take hold." The answer, for starters, is that "subsystems can optimize local goals, while 'carelessly' undermining system goals." We begin to see the solution, one that encompasses the need for subsystems to interact laterally – a conclusion essentially similar to our own work in fostering "deep learning" across large school systems (Fullan, Quinn, & McEachen, 2018). Of course, there is much more to what needs to be done, which you will find in the chapter. The strategy involves greater heterogeneity of interaction enabled by leaders who can help forge new patterns of learning that apply to larger understandings, and new solutions.

Chapter 2 applies systems theory to the turnaround school problem. Systems theory contains the distinction between single loop and double loop learning which always sounds mysterious. We can simplify it here. Single loop involves one variable, while double involves two. The full possible model is: theory > behavior > consequence. Most (non-systemic) strategies involve single factors such as: expecting new behavior, like practicing certain reading techniques, would result in improved reading. This is single loop learning (expecting one factor to make a difference). But let's say we engage in double loop learning, where we add teacher beliefs, such as, does the teacher believe that the student is capable? It turns out that the teacher does not believe the student is capable, and when the behavior is not forthcoming they have an "I told you so" attitude. No improvement ensues and the teacher is proven right in the circumstance. Now systems theory comes into play. The teacher is helped to question her beliefs and is paired with another teacher who demonstrates that high expectations combined with behavior actually works. We don't have to call it systems theory but that is what it is: behavior (single loop) doesn't work, but if backed up by discussing and questioning beliefs (double loop) that turns out to make a difference. The conclusion, which the principal discusses, is that if teachers learn together, examine expectations, discuss their own beliefs, and test them out, then the group can make a difference even in difficult situations. What they do is take one more factor into account, namely, the teacher's belief. This is systems theory at work.

The above-mentioned example is at the individual school level. We can seek more systems-level confirmation if we turn to an example from Part III of the book (no need to be linear) where turnaround schools are considered for a whole state – Victoria, Australia. Here systems theory takes us to consider a state-wide approach that involves learning and improvement cycles where schools within the system learn from each other. The third chapter at the school level focuses on equity, and shows that treating equity as a system (as distinct from a technical problem) can make a world of difference. The authors make the case that complex social problems require getting at the views of those who may not be at the table, and are the ones experiencing the problem.

The final chapter of Part I – still at the local level – focuses on how we might measure systems thinking in elementary school principals. This chapter not only identifies the components of school principals' systems thinking, but also shows its significant correlations with instructional leadership and organizational commitment.

Part II of the book contains four chapters at the middle level of the system (at the district or regional level). Chapter 5 compares system growth models in two states: Ohio and Texas. The chapter gets at critical differences between the two states, and more discussion on the causes or reasons for the differences could be a valid topic for future study.

Chapter 6 presents a wonderful analysis of how systems theory was used to cultivate changes from within, relative to establishing a culture of sustainability within a whole district. Diverse networks were combined with high levels of trust to establish and culture green-based sustainability. Eight critical incidents are identified that constitute an effective theory of action for district sustainability. We see a combination of front-end communication and focused relationship building that permeates the whole system.

Chapters 7 and 8 complete the cases at the district or regional levels. In Chapter 7 we see an aspect of district change that is often neglected, namely the relationship between the school board and the superintendency. As we see in our own recent work, sustainable change requires developing rapport between school trustees and superintendents (Campbell & Fullan, forthcoming). For systems change, district leaders must frame the discussion in interaction with wider evidence related not just to educators but also to community, business leaders, and related multiple stakeholders. In Chapter 8, "System Leadership for Continuous Improvement," we see a model that is explicitly focused on developing system solutions consisting of: system outcomes, variation, system perspectives, evidence-based processes and outcomes, and involving specific mental models, behaviors, and focuses. This chapter shows how system leaders think, and what actions they take to foster a culture of system improvement.

The final section of the book contains four chapters on systems thinking at the state and national levels. Chapter 9 examines the impact of social media on actions in complex systems. Chapter 10 shows the relationship between whole system action, and local action that reveals both their mutual autonomy and their interrelationships. Chapter 11 focuses on school turnaround across the whole system using system theory as both an analytic tool and a strategy of school improvement. Systemic thinking reveals a four-fold table of system interaction aimed at turnaround based on type of interaction: confluence, collision, pull, and push. The last chapter once again takes up the question of turnaround schools across a whole state (Victoria, Australia). This chapter reveals the importance of and need for combining system leadership and school leadership for whole system reform.

All in all, *Leading Holistically* is a one-of-a-kind book. It provides a set of examples of what leading systemically looks like across the three levels of systems. It gives us "applied system thinking in action." It sets a new standard for systems theory. Above all it invites new applied scholarship using the power of systems thinking to understand and take actions that affect whole systems for the better. It presents a new and powerful friend into the neighborhood. It makes system thinking that much more accessible.

## References

Boulding, K. E. (1956). General systems theory: The skeleton of science. *Management Science*, *2*, 197–208.

Campbell, D., & Fullan, M. (forthcoming). *Coherence and local governance*. Thousand Oaks, CA: Corwin Press.

Fullan, M., Quinn, J., & McEachen, J. (2018). *Deep learning: Engage the world change the world*. Thousand Oaks, CA: Corwin Press.

# PREFACE

## The Systems Language

The rapid changes and increased complexity of today's world present us with a new and challenging phase for education systems worldwide, requiring fresh thinking and new approaches. Simply put, the traditional tools at the disposal of policymakers, district leaders, and school principals facing today's educational leadership complexities are insufficient. Educators sorely need comprehensive frameworks accompanied by complementary strategies to guide them toward success. We believe that systems thinking may be extremely beneficial for dealing with the educational challenges of the twenty-first century.

To explain systems thinking in general terms, we will begin with the opposite point of view, which is beautifully expressed in the famous parable about a group of blind men who wanted to explain an elephant, by feeling various parts of its body. When they conceptualized the elephant based on their incomplete experience, their descriptions were fundamentally different. A poem based on this Indian parable, written a hundred and fifty years ago by John Godfrey Saxe, concludes: "And so these men of Indostan / Disputed loud and long, / Each in his own opinion / Exceeding stiff and strong, / Though each was partly in the right, / And all were in the wrong" (Saxe, 1873, pp. 77–78).

In contrast to how these blind men chose to go about grasping the issue at hand, systems thinking is about seeing the whole picture. It advocates stepping back and focusing not only on the trees but also on the forest – with the trees symbolizing particular situations or limited domains, while the forest symbolizes the overall view. Instead of being overly focused on discrete details, systems thinking urges its practitioners to discern an overall pattern from a collection of details, aggregating and integrating this data into a holistic framework. Systems thinkers see a broad, general view of the system's components and environment, and direct their efforts accordingly. Therefore, systems thinkers are able to understand the system both conceptually and functionally, even without understanding all its minutiae. They seek not only tactics, which are incremental steps and short timeframes along the way, but also a strategy, which is oriented toward achieving broader missions and long-term goals.

The ancient Greek philosopher Aristotle laid one of the foundations of systems thinking by positing that a whole is greater than the sum of its parts. From this perspective, the elephant is not only a combination of trunk, tusks, ears, legs, and tail; in fact, the elephant as a whole lends meaning to the trunk, tusks, ears, legs, and tail. Along the same lines, the only way to fully understand a system is to understand its parts in relation to the whole, because it is the whole that endows the parts with meaning. Moreover, the elephant's defining characteristics indeed comprise the characteristics of the elephant in its entirety. The defining characteristics of the whole system cannot be found in the isolated parts, because once the system is analyzed – i.e., taken apart – these defining characteristics lose at least some of their meanings.

The elephant is not the sum of its parts; it is a product of the parts' interactions. Systems thinking is a holistic perspective that focuses on how the parts function together in networks of interaction, not on breaking down systems into parts in order to understand them separately. The parts of a system operate in intimate interconnections, with the interactions among them creating the whole and shaping its defining features. Systems thinking provides a means of seeing the system as an integrated, complex composition of many interconnected components that need to work together for the whole to achieve success. Thus, to improve the whole, systems thinkers optimize the interactions among the system's parts. Interaction management may result in improved performance, reduced conflicts, expanded delegation of responsibility, and overcoming of resistance.

The figurative elephant discussed in this book refers to education systems. This text's goal is to explore educational leadership and policy through the systems thinking lens, placing the study of wholes before that of parts. Research on systems thinking in the context of educational leadership and policy is scarce, although its potential contribution to various educational issues has received growing attention in recent years. Aiming to expand the existing literature on systems thinking in educational leadership and policy, this book deals with three concentric spheres: the individual school level, the district/regional level, and the state/national level.

Thus, this book is divided into three main parts: systems thinking at the individual school level, systems thinking at the district/regional level, and systems thinking at the state/national level. Most importantly, the major theme focuses on how these three concentric spheres can be woven into a coherent whole. Despite this analytic division, the chapters examine how the three different levels of systems thinking in educational leadership are interrelated and interconnected and in effect are the system.

The first part of this book focuses on systems thinking at the school building level. Therefore, it deals with systemic aspects of school leaders' practices in general, and within complex contexts of diversity, low-performing schools, and educational reform implementation in particular. Despite their focus on the individual school level, the chapters in this part broadly consider the school environment, carefully examining the school's interactions with the district and national levels.

In Chapter 1, Ebony N. Bridwell-Mitchell (Harvard University) conceptualizes school organizations as systems, exploring the implications of this perspective on understanding how schools respond to state-mandated reforms. An illustrative case shows that school responses are driven, at least to some extent, by socio-cognitive control processes, which under identified conditions maintain the status quo. These conditions stem from highly differentiated informal school subsystems, suggesting that an inability to coordinate competing subgoals across subsystems had contributed to the school's ultimate closure.

In Chapter 2, Sharon F. Rallis (University of Massachusetts Amherst) and Rachael B. Lawrence (University of Massachusetts Amherst) investigate a failing urban elementary school that has demonstrated a holistic change directed by systems thinking, describing how this approach applied by key actors in the school and district led to dramatic changes in the school's teaching and learning culture. To this end, the chapter focuses on systemic interactions, perspectives and boundaries, providing insights regarding the courage, risks, and actions that schools should apply to address the interconnectedness of human interaction to lead system change.

In Chapter 3, Sharon D. Kruse (Washington State University) demonstrates how school leaders can benefit from the soft systems thinking perspective, illustrating how it may be applied in addressing issues of equity within schools and school districts. Equity challenges do not exist independent of school leaders, teachers, or students and their families; they are deeply contextual and situational, and are also open to interpretation. Surfacing different views, including those related to problem-definition and potential explanations or decisions, is a hallmark of the soft systems approach, and can be utilized to cultivate equity leadership in schools.

In Chapter 4, Haim Shaked (Hemdat Hadarom College of Education), Pascale Benoliel (Bar Ilan University), Nechama Nadav (Bar Ilan University), and Chen Schechter (Bar Ilan University) present an instrument developed to measure systems thinking in elementary school principals. Their findings indicate that principals' systems thinking comprises the following factors: evaluating significance; openness to a variety of opinions; leading wholes; and adopting a multidimensional view. In addition, principals' systems thinking was found to correlate significantly with two relevant established constructs: instructional leadership and organizational commitment.

The second part of this book expands the horizons being viewed, discussing systems thinking at the district level. In this part, systems thinking is used to explore topics such as cross-district results and whole-district transformations. Naturally, the superintendent's role receives special attention in this part, throughout which the district is viewed as having a reciprocal relationship with both the individual school level and the national/federal level.

In Chapter 5, Alex J. Bowers (Teachers College, Columbia University), Xinyu Ni (Teachers College, Columbia University), and Jennifer Esswein (Education Northwest) apply hierarchical linear growth modeling to district demographic and effectiveness data in order to identify districts that are significantly outperforming or underperforming their demographic and resource characteristics compared to the entire population of other districts. They found that multiple district demographic variables and financial expenditures have a significant connection to growth in the district level's academic achievements.

In Chapter 6, Cynthia L. Uline (San Diego State University) and Lisa A.W. Kensler (Auburn University) employ instrumental case-study methods to gain an in-depth understanding of a system-wide reform within a large California urban school district, aimed at "greening" the district from the inside out. Findings show how district and school level leaders applied systems thinking to invite creative solutions and adaptations, building a diverse network of support for sustainability-related building, management, and curricular and instructional practices throughout the district.

In Chapter 7, James Coviello (Saint Joseph's University) and David E. DeMatthews (University of Texas at Austin) provide a framework for examining the functioning of superintendents who aspire to achieve equity-oriented leadership. Aiming to influence the process of sense-making and to inspire action, these superintendents effectively frame problems and propose solutions with messages that draw on the ideas and language that already exist in the broader community. This model of holistic engagement focuses on building broad coalitions and influencing various internal and external stakeholders.

In Chapter 8, Christina J. Dixon (Carnegie Foundation for the Advancement of Teaching) and David Eddy-Spicer (University of Virginia) examine the current research-based knowledge at the intersection of the fields of quality improvement, education, and leadership. Based on the existing literature, which highlights how successful leaders of continuous improvement think, what they do, and where they focus their efforts in order to address systemic challenges, the chapter presents an initial model depicting how educational leaders initiate and sustain systemic continuous improvement across their districts.

Encompassing the whole picture, the third part of this book engages in systems thinking at the national level. It thus engages in topics such as educational policy setting, state-wide programs, and the system of social media which swirls around and interacts with educators. Overall, systems leadership in education may be seen as funnel-shaped, starting with the national school system as a whole, going through the district, and ending at the specific destination of the individual school.

In Chapter 9, Alan J. Daly (University of California, San Diego), Miguel del Fresno (Universidad Nacional de Educacion a Distancia, Madrid, Spain), and Jonathan A. Supovitz (University of Pennsylvania) seek to understand the meaning-making that occurs within the growing social media system around educational policy and systems leadership. Relying on data from a recently completed longitudinal study of nearly 200,000 actors, they offer insights into the impact of transactions among "socially influential" actors over time in this new social media/educational policy space, expanding traditional concepts of the "system" in systems-level leadership.

In Chapter 10, Adam Nir (Hebrew University of Jerusalem) proposes systems thinking as a useful framework for policy-setting processes where policy makers face dynamic and complex circumstances which involve heterogeneity and contradictions. Based on the analysis of an Israeli policy program, the chapter offers a typology of the compromises made by policy makers, suggesting that determining reasonable boundaries for systems thinking processes performed in loosely coupled systems is essential if these processes are to assist policy makers rather than increase the complexities confronted by them.

In Chapter 11, Andrea Rorrer (University of Utah), Vicki Park (San Diego State University), Cori Groth (University of Utah), and Janice Bradley (University of Utah) apply a systems theory approach – the Confluence of Influence theory – to examine the evolvement of a state-wide leadership support program for turnaround leadership. They discuss how this policy and its funding context have created the "space" and intentionality for the program to develop, and how the various actors within the systems played various roles in negotiating and developing the program.

In Chapter 12, David Gurr (University of Melbourne) and Lawrie Drysdale (University of Melbourne) integrate a variety of research sources on system leadership in the state of Victoria, Australia at the system, regional, and individual school levels. Based on these pieces of interrelated research, they claim that whilst system leadership can be helpful at the individual school level, the work of the principal, other school leaders and critical friends was more important to the school's improvement journey. Without effective school leadership, system leadership's impact is likely to be limited.

Moving beyond the typical focus on discrete units, this book may usher in a new epoch of systems-level leadership and focus. Too often, school leaders and policy makers are peddled simplistic solutions to complex situations. As so many leaders have discovered, these panaceas unfortunately rarely work, because they are not sufficiently holistic. By focusing on parts rather than on the whole, and by assuming that there is one best solution that fits all circumstances, educational leaders miss the opportunity to adequately consider the complex interactions that occur among various parts of the system. Hence, without systems thinking, educational leaders may find it nearly impossible to face contemporary growing complexity, change, and diversity. The current book suggests a systems-thinking approach that integrates both theory and practice for educational leadership at the state, district, and school levels. It reframes the idea of educational change within a broader idea, i.e., the systemic one. This idea has significant implications on how we conceptualize the "elephant" of educational change and the actions to be taken in order to realize the promise of improvement for all those connected to and involved in the important endeavor of education.

## Reference

Saxe, J. G. (1873). *The poems of John Godfrey Saxe: Complete edition*. Boston, MA: James R. Osgood and Co.

# PART I

# Systems Thinking at the School Level

# 1

# SYSTEM SELF-REGULATION AND STATIC EQUILIBRIA

## How Socio-Cognitive Control Processes Regulate School Responses to External Reform

*Ebony N. Bridwell-Mitchell*

The phrase "school system" commonly refers to the set of schools and related services or supports under the jurisdiction of a particular governing body, such as a district, board of education, or state education agency. This usage is not incorrect. Nevertheless, this usage focusing on the *elements* comprising the system – schools, services, supports – obscures what is arguably the key implication of the phrase – the existence of particular *dynamics*, which make schools a "system" and, in fact, make each individual school in that system a (sub)system unto itself.

A system, particularly a complex adaptive system, such as a school, is a set of numerous, diverse, interdependent elements whose intentional and patterned interactions generate aggregate behaviors, which are greater than the sum of their parts (Boulding, 1956; Holland, 1995; Morel & Ramanujam, 1999). In much the same way that the laws of physics determine the operation of the physical universe, school operations and outcomes are determined, in many ways, by the fundamental principles of systems. This means that rather than seeing phenomena in schools as idiosyncratic, irregular, or unexpected – and thus difficult to explain, predict, or address – many issues in schools can be understood as the result of system dynamics. One example is the persistent challenge schools face when implementing reforms, such as instructional improvements in classrooms (Payne, 2008).

In this chapter, I leverage a systems perspective to explore how self-regulating routines, which tend to maintain a state of homeostasis or relative equilibrium in a system, help explain why state-mandated instructional reforms failed to take hold at one school (Ashmos & Huber, 1987; Boulding, 1956; Kast & Rosenzweig, 1972). I use this case as an illustration of the utility of the systems perspective for illuminating the dynamics of complex phenomena in schools and making sense of outcomes, which appear to run counter to desired goals, such as school improvement. Specifically, I argue and demonstrate that school organizations' responses to external shocks, such as state-mandated instructional reforms, are driven by self-regulating control processes of socio-cognitive systems, namely (1) social comparisons to social standards, (2) feedback through observation and conversation, and (3) adjustments in understandings, attitudes and practices via social influence.

Before presenting the case, I begin, in the next section, with an overview of the systems perspective. This is followed by a set of arguments about the implications of the perspective for understanding system responses to external shocks, such as mandated instructional reform. Because this chapter is intended to illustrate the systems perspective, I formalize the arguments by employing a method of qualitative modeling (i.e., Heise & Durig, 2000), which reflects one type of formal logic on which systems thinking can rely. In the discussion, I point to the important role of system structure and conclude briefly by emphasizing the utility of systems theory for educational researchers and practitioners.

## Understanding System Principles

The systems approach, which is taken up across many different areas of science, highlights the key role of connectivity, chains of cause and effect, change over time, and patterned behaviors or structure (Boulding, 1956; Von Bertalanfy, 1972). In the social sciences, the perspective has taken root most firmly in the study of organizations (Katz & Kahn, 1978; Miller, 1972; Scott & Davis, 2007). Thus, the system perspective is of special importance for understanding schools as organizations (i.e., Bidwell, 1965). In the study of organizations, the systems perspective became prominent through the field-defining work of Katz and Kahn (1978). These two social psychologists sought to understand how the complex interplay of human behavior affected organizational outcomes. To do so, their work stressed the role of certain principles, which were the same principles that could be observed in so-called *open systems*, which depend on throughputs or exchanges (i.e., of energy, information, resources) with the surrounding environment.

One of these system principles is *negative entropy*, meaning a tendency, over time, toward increasing order and stability given sufficient inputs from the environment (Ashmos & Huber, 1987; Kast & Rosenzweig, 1972; Katz & Kahn, 1978). Think, for example, of the rigidity of certain structures and practices in older organizations or among veteran members of organizations. The path to increasing order unfolds through the principle of *differentiation* (Ashmos & Huber, 1987). This means that complex activities are broken down and elaborated into smaller sets or subsystems of activities and functions. In organizations, the result tends to be adding new units or roles responsible for newly identified organizational sub-goals. An example might be adding an assistant dean of school culture, office of family engagement, or a teacher team charged with implementing a new math curriculum.

While the principle of differentiation suggests constant growth and change in organizations, another principle, *homeostasis*, means systems tend toward an equal balance between competing forces or influences. The result is a relative degree of certainty, control, coherence, and harmony, the preference for which is one reason there often is aversion to change in organizations (Ashmos & Huber, 1987; Katz & Kahn, 1978). Ultimately, homeostasis is achieved through a final system principle: the existence of *self-regulating feedback routines.* Self-regulating routines provide information to the system about the extent to which it is achieving desired goals given external demands, environmental inputs, and relative to an internally balanced state – i.e., homeostasis (Ashmos & Huber, 1987; Kast & Rosenzweig, 1972). As described in a subsequent section, self-regulating feedback routines play an important role in how organizations respond to environmental shocks, such as how schools respond to external reform mandates.

## Self-Regulation in Socio-Cognitive Systems

### *How Control Processes Create and Restore Balance in Systems*

One principle of systems is that they are self-regulating, meaning they adapt to perturbations or shocks through processes that return the system to a state of equilibrium or balance, which may reflect new local optima given changed environmental conditions (Ashmos & Huber, 1987; Forrester, 1968; Kast & Rosenzweig, 1972). Some of the earliest thinking about systems, sometimes referred to as *general systems theory*, proposed that system self-regulation was determined by a set of identifiable control processes (Boulding, 1956; Von Bertalanfy, 1972). These control process are comprised of three routines: (1) comparisons to a standard, (2) feedback about deviations, and (3) required adjustments.

These self-regulating routines are easily conceptualized in simple, cybernetic systems, such as a thermostat (i.e., Scott, 1998). In the case of a thermostat, a standard is set as the goal for system behaviors and outputs, namely the desired room temperature. System behaviors and outputs are regularly monitored and compared to the standard – whether the room is above or below the desired temperature; whether the furnace is off or on. These comparisons provide feedback or information to the thermostat about required functions. When this feedback indicates that there is sufficient deviation from the standard or goal, the result is an adjustment in system behaviors and outputs. In other words, the furnace turns on or off in order to achieve the desired temperature (Forrester, 1968; Scott, 1998).

Organizations are clearly more complex than a simple system, such as a thermostat. Indeed, as the systems perspective has evolved, organizations have come to be understood as complex adaptive systems, which operate in less predictable and sometimes chaotic ways, including the possibility of having multiple, episodic equilibria of which homeostasis is only one (Amaral & Uzzi, 2007; Frank & Fahrbach, 1999; Maguire, McKelvey, Mirabeau, & Oztas, 2006). Still, systems theorists argue that the basic features of lower level systems tend to be incorporated into systems at successively higher levels (Ashmos & Huber, 1987; Boulding, 1956; Scott, 1998). This means that while control process in complex adaptive systems might appear to be very different from that of simple systems, they nonetheless exist and operate to affect system dynamics and outcomes. The question is how principles from lower level systems translate to higher level systems. In other words, what might control processes in a complex adaptive system such as an organization, and a school organization in particular, look like?

## *Understanding Socio-Cognitive Control Processes*

Consider that organizations, as systems, are social and technical systems, in contrast to simpler cybernetic or mechanical systems (Ashmos & Huber, 1987). Rather than wires, chips, or gears, the components of organizations are human agents along with the tools and technologies upon which they rely to perform their work. This includes, for example, tools such as particular governance structures or digital learning platforms and technologies such as certain information systems or pedagogical practices (Scott & Davis, 2007). However, one important aspect of the role of material objects in organizations is not simply that they exist but that actors make sense of material objects and use them to enact and perform the work of the organization (c.f. Callon, 2009; Fenwick & Edwards, 2010; Latour, 1991). This suggests the importance not only of social actors in organizations but also the importance of their cognitions – about the tools and technologies of their work but also about organizational goals and the broader environment in which the organization is situated (Spillane, Reiser, & Reimer, 2002). This is to say that rather than simply being socio-technical systems, organizations are socio-cognitive systems (Dunn & Ginsberg, 1986).

If organizations are socio-cognitive systems, then the control process that regulate organizations are most likely cognitive and social routines. Instead of a predetermined standard, like the temperature set for a thermostat, the standard in a socio-cognitive system is more likely a social standard or set of norms, which emerges from the interactions and shared understandings of organization members (Fitzpatrick, 1989; Friedkin, 2001; Sherif, 1966). Likewise, comparisons that determine deviation from the standard would be social comparisons, meaning individuals turning to peers to assess the appropriateness of their beliefs and behaviors in order to minimize uncertainty about the accuracy of their own self-evaluations (Festinger, 1954; Merton & Rossi, 1957). As a part of control process in organizations, this would mean that one way organizational behavior is regulated in response to an external shock, is for organizational members to turn to peers and colleagues to determine how they should react (e.g., Coburn, 2001). In this case, how to react means how their existing understandings, attitudes, and work activities should be adjusted, if at all.

If standards and comparisons in socio-cognitive systems are social, then feedback in the control processes is also very likely social. Two forms of feedback which are social in nature are observations (i.e., of others) and conversations (i.e., with others). So, with social feedback, information about system performance is provided by organizations members' observations of others' work activities and goal achievement and by conversations about the same. In this way, conversations and language along with organization members' interpretations of them are mechanisms for system self-regulation and are a potential force for organizational constancy or change (Ford & Ford, 1995; Hill, 2001).

To the extent that organization members do make changes in their understandings, attitudes, and work activities, this results, in part, from the lessons learned from peers through observation and conversation – namely that there is a need to make adjustments in order to match the social standard set by peers. This kind of change and adjustment across actors in a social setting matches almost exactly the described process of social influence, which occurs within and across subgroups of actors (Fitzpatrick, 1989; Friedkin, 2001; Katz & Lazarsfeld, 1966; Moore, 1968). In other words, the final control process or self-regulating mechanism in organizations, as socio-cognitive systems, is social influence, which determines adjustments made in the system.

## *A Model for Socio-Cognitive System Responses to Exogenous Shocks*

To summarize, one way to understand how organizations respond to external shocks, such as how schools respond to mandated instructional reforms, is to examine the self-regulating control process of socio-cognitive systems. This is namely the role of comparison to social standards, feedback through observation and conversation, and adjustments via social influence. One way to make the arguments about socio-cognitive system responses more concrete is to formalize the proposed model. In contrast to a mathematical formalization, I formalize the arguments with a method of qualitative modeling (i.e., Heise & Durig, 2000). The model uses natural language operators (see Table 1.1) to present formal statements of first-order logic – meaning statements about causal chains between sets of variables – about socio-cognitive responses to external shocks (see Hannan, 1998; Pollock, 2002, for examples; see Tarski, 1946, for an introduction).

It should be noted, however, that while the formal model is useful for making the above arguments more concrete and for illustrating the underlying logic of the arguments, the model should not be taken as a perfect representation

**TABLE 1.1** Natural Logic Operators and Variables for First Order Logic Statements

| *Logic Operator* | *Statement* | *Variable* | *Domain Construct* |
|---|---|---|---|
| ∀ | given any | ED | External Demands |
| ∃ | there exists | UN | Uncertainty |
| ⊃ | if-then | SC | Social Comparison |
| → | implies | FB | Feedback |
| = | corresponds | OP | Observed practice |
| ≠ | does not correspond | CP | Conversations about practice |
| / | given | SS | Social Standard |
| ∧ | and | AJ | Adjustment |
| \ | or | | |
| ~ | not | | |

of the system dynamics. This is because the formal model, by relying on first-order logic statements denoting causal chains between variables, represents the relationships of interest as linear when the dynamics of complex adaptive systems tend to be non-linear (Amaral & Uzzi, 2007; Holland, 1995; Rivkin & Siggelkow, 2007). Still, the formal model is helpful as a framework against which descriptive evidence can be compared and so aids in being able to falsify system dynamics arguments, which some have argued is a limitation of the perspective (Maguire et al., 2006). Thus, the findings from the presented case used to illustrate arguments about socio-cognitive control processes should be evaluated against the statements in the following model:

**MODEL 1.** Socio-Cognitive Control Processes for System Responses to Exogenous Shocks

Statement 1: GIVEN ANY external demand requiring new work functions (∀ED)

1.1 THERE EXISTS uncertainty about standards for work functions (∃UN)

Statement 2: GIVEN uncertainty actors employ social comparisons to determine standards (SC/UN)

Statement 3: Social comparison IMPLIES actors receive feedback GIVEN observed practices AND conversations about practice (SC → FB/OP ∧ CP)

Statement 4: IF feedback DOES NOT CORRESPOND with the social standard THEN adjustments via social influence result (FB ≠ SS ⊃ AJ)

4.1 IF feedback CORRESPONDS with the social standard THEN adjustments via social influence do NOT result (FB = SS ~ AJ)

## The Case of Embark Middle School[1]

### *The Requirements of State-Mandated Reform [Statement 1: ∀ED; Statement 1.1; ∃UN]*

Embark Middle School (Embark) sits two blocks south and east of one of the busiest thoroughfares in one of the large northeastern city's most populous neighborhoods. Like most schools undergoing mandated state reform, Embark serves mostly poor students of color and has had a history of underperformance. In the year prior to reform, 25.1 percent of students at Embark were Black, 71.5 percent were Hispanic, and 2.4 percent were Asian or another racial-ethnic group; 97 percent were eligible for free or reduced lunch; and 93 percent of the students were below grade level in English language arts and mathematics. The school has four administrators and 68 teachers assigned to a variety of subject specialties. As described by a teacher's aide who had been working at Embark for 20 years, the school had a tumultuous history: "The [city newspaper] listed us as the worst middle school in the whole city. There were fires, throwing things out of the windows, and kids selling drugs . . ."

The main aim of state reform was to have teachers structure their lessons using a practice termed "the workshop model." In contrast to a traditional model in which the teacher lectures in front of the classroom to passive students sitting quietly in rows, the workshop model entails students first spending brief periods gathered at the front of the room listening to the teacher present specific content, and then moving to tables to work independently or in small groups informally supervised by the teacher. As part of the workshop model, teachers are expected to differentiate instruction and incorporate a variety of materials that meet students' diverse learning needs.

Embark teachers were uncertain about using the workshop model, in part, because they doubted its efficacy, as one Embark teacher, who had been teaching English language learners at Embark for three years, explained:

> The workshop model does have its steps, but if a child can't really write you have to plan ahead. Maybe he could draw or he can put down some words . . . The workshop model is not for everybody. That program is just for kids who are in the upper grades, and even they can't read well; they're really below grade level.

Other teachers doubted the new instructional schedule put into place with the workshop model, its pace, structure, and the breadth of new practices. One teacher lamented, "We have five minutes to do the read-aloud. They [students] have independent reading; it's supposed to be 20 minutes. Everything has a time when it is finished; if the time is out, then you won't get to the end of the lesson . . ." Another teacher explained her doubts as follows:

> Now, there is so much specific material that needs to be covered . . . because before, if kids got something, you could just jump to something else that was outside of the curriculum but would be more interesting. Now, so much emphasis is placed on following the curriculum map that there's not much room for anything else.

Even with their doubts about the efficacy of the workshop model and their confusion about how to implement it, Embark teachers also doubted their past practices given the state's condemnation. These kinds of concerns were pervasive among Embark teachers but were explained succinctly by one sixth-grade teacher who reflected on his own past instructional practices this way:

> Maybe a lot of the changes that are happening were needed. Some of them are tough to take. It's hard to look at yourself and think that you weren't doing things the best you could have been doing . . . It kind of humbles you a little to know that you know you have to do a better job . . . to know you have to constantly be trained.

Embark teachers' doubts about their past practice, about the efficacy of the workshop model, and about their ability to implement it, problematized teachers' day-to-day classroom practice. In order to make decisions about classroom practice, Embark teachers would first have to resolve their instructional dilemmas. Embark teachers tended to resolve their uncertainty and instructional dilemmas by first trying to determine what the appropriate instructional standard should be, which they did by making comparisons to their colleagues.

### *Social Comparisons to Colleagues' Social Standard [Statement 2: SC/UN]*

Embark teachers resolved dilemmas about which practices to use and how by first trying to learn about their colleagues' instructional aims, attitudes, and activities. In other words, colleagues' instructional aims, attitudes, and activities provided a social standard or norm that helped teachers understand what they were supposed to do. In order to learn the social standard, Embark teachers made frequent and ongoing comparisons to their colleagues starting at the very beginning of the school day.

For example, Embark teachers used colleagues' arrival and departure times to judge their preparation and commitment to instruction. Mr. Mackey, a seventh-grade social studies teacher, explained this by saying, "So I mean, if you come into school at 8:30 and you leave at 3:00, and you have no plans, I think that says something . . . it says something to me, I make judgments; even if you don't want, we always make judgments." The judgments to which Mr. Mackey[2] referred are judgments about the apparent norm for instruction, given the implications of many teachers not coming in early enough to prepare their classrooms or not staying late enough to work with students who need additional help.

Observations of peers and comparisons to the social standard are a natural part of most social contexts; however, comparisons become particularly pervasive when individuals are faced with uncertain or ambiguous situations (Festinger, 1954). Thus, the instructional dilemmas of reform at Embark made collegial comparisons the subtext for ongoing collegial interactions. These interactions may have had other, more explicit functions, yet teachers took these opportunities to compare their instructional aims, attitudes, and activities with those of their colleagues.

Comparisons to colleagues occurred when Embark teachers congregated during breaks to have coffee, celebrated birthdays, or took short walks to the corner store and had conversations about the lessons they had taught recently. Ms. Mason, a sixth-grade teacher, explained how teachers relied on collegial comparison to determine appropriate instructional practices by recalling a recent visit to a colleague's classroom. She noted that when visiting a colleague's classroom, "You come back and come away from that room, well you know, [you say,] 'I should try that in my room' or 'Can I try that in my room? I don't think my kids are up to speed yet, but maybe I can tone it down and have them be responsible for something else.'"

Embark administrators also were very clear about the utility of comparisons among teachers. Indeed, administrators often engineered opportunities for teachers to make comparisons. This was illustrated by Embark's assistant principal, who explained how she motivated some of her teachers to adopt practices prescribed by reform:

> I think that peer, collegial conversation is more important than myself or the other APs going in. We do have teachers that have difficulties, and I'll say, "Well, Ms. So and So, why don't you go and talk to Mr. So and So and see how that goes?" That works best. I believe that the large group is the biggest influence on the small group.

### *Informal Peer Feedback via Observation and Conversation [Statement 3: SC $\rightarrow$ FB/OP $\wedge$ CP]*

Collegial comparisons provided guidance to Embark teachers because they used comparisons to gather information about their colleagues' instructional aims, attitudes, and activities. In addition to gathering information, Embark teachers also used collegial comparisons to determine how much their own instructional aims, attitudes, and activities differed from those of their colleagues. Information about this distance between their instructional approaches and those of colleagues and about how to close the gap constituted informal peer feedback about instruction.

One Embark teacher explained how teachers relied on peer feedback by saying, "you can get together with any teacher in the school any time, you know, share ideas about lessons, to talk about students that you both teach, so that's helpful [for teaching] . . . ." Another teacher described the regular meetings she and her colleagues had to share information about instruction: "We meet weekly, every Monday to discuss issues or just talk over things that we need to know, I think just touching base with all our teachers, I think this is helpful."

When compared to feedback from administrators, peer feedback involved higher levels of trust and occurred between individuals with similar organizational experiences. As a result, peer feedback had a strong influence on teachers' understandings of practice. This was previously explained above by Embark's AP when she noted that "the large group is the biggest influence on the small group." As another example, while describing why some teachers changed their instructional practice, this Embark teacher explained:

> I really believe it's those informal observations that we do in the classroom [that influences teachers]. It gives us the opportunity to discuss what we saw, privately with each other. It's not about "I got you!" It's not about being evaluated as satisfactory or unsatisfactory. It's about seeing what other teachers are doing. Those private conversations about what we saw; those are very helpful.

This same sentiment was expressed by another Embark teacher emphasizing the importance of informal observations of colleagues: "I think the best way to learn is to actually observe. Through observation, I mean, you kind of get some idea [of what to do], although you may not get the full picture. But, I think speaking to other teachers, you know, is a good way to learn what you're supposed to do."

While receipt of feedback was sometimes purposeful, it was also incidental. This happened, for example, when Embark teachers observed the artifacts of colleagues' instruction, which were present in classrooms even when no instruction was occurring. Teacher-created charts left posted on a blackboard revealed students were working on an interdisciplinary, theme-based project. Desks arranged in small groups instead of in rows indicated that students were working in cooperative groups. Classroom materials, such as books, posters, and manipulatives indicated whether diverse learning styles and cultural backgrounds were being attended. Ms. Kellogg, an eighth-grade social studies and reading teacher, explained how repeated feedback from colleagues helped teachers reflect on their instruction and think about how to improve:

> You know, if teachers hear from others what they are doing, if they see what everyone else is doing and it's different from what they're doing then they start to reflect. You know? That makes you look at what your

> accomplishments are, look back on your goals . . . ; [that difference makes you] ask what needs to improve.

An exchange between two Embark teachers who stopped on the third floor in front of a bulletin board also illustrated how collegial feedback helped teachers determine how their personal instructional approaches deviated from those of their colleagues. One teacher, scrutinizing a posted paper, said to his colleague, "I see you gave that student bad marks." The second teacher, who was quiet for a moment, while smoothing down the paper's crumpled edges, replied, "I might take that down and let him do it again. I'm not sure that's really his best work." The first teacher's reply, "Mmmm . . . hmmmm," vocalized in a slow, questioning tone, signaled his apparent doubt or reproach and let his colleague know that her standards for the quality of work posted on the bulletin board might not have matched his own. The next day, the paper in question had been removed and replaced with another student's work.

The feedback exchanged between the above two teachers reinforces how exchanging peer feedback is an important routine for generating shared understandings. First, feedback spreads or diffuses information about appropriate instruction from teacher to teacher so that internal meanings become externalized messages. As information spreads, teachers construct understandings about instruction based on meanings implied by their colleagues' instructional attitudes, aims, and activities. Teachers who receive similar messages, construct similar meanings and understanding of instruction. Peer feedback at Embark also encouraged shared understandings by acting as a conduit for peer pressure or influence to conform to colleagues' understandings of instruction.

## *Peer Influence and Updating [Statement 4: $FB \neq SS \supset AJ$; $FB = SS \supset \sim AJ$]*

When Embark teachers explained what it would take for their school to change they gave explanations like this one given by Mr. Ramon: "So everyone should be on the same page . . . some of that old stuff [of how we used to teach] creeps in but really we just all have to get on the same page." Similarly, Ms. Branch explained, "I guess, the most important thing is that the higher-ups, really everybody is on the same page." In order to get on the same page, Embark teachers' potentially distinct understandings of reform requirements had to become less distinct and more similar to those of their colleagues. Distinct understandings among Embark teachers became more similar as they updated their instructional understandings to match those of their colleagues. This updating was the result of Embark teachers exercising and responding to peer pressure or social influence from their colleagues.

Influence and updating was described above by the teacher who explained how seeing her colleagues' classrooms made her ask if she could do something similar in her own classroom. The process of influence and updating also was explained by the teacher who said that seeing what everyone else was doing made her reflect on how she could improve and by the teacher who replaced student work on the hallway bulletin board after comments made by a colleague. Thus, at Embark, influence and updating was a consequence of collegial comparisons and informal peer feedback. Taking a closer look at comparisons and feedback among Embark teachers reveals that influence unfolded in a number of specific ways. This included the use of shared symbols or language, the impact of superior numbers, providing social sanctions and rewards for conforming to peers' expectations, and offering explicit directions about which instructional aims, attitudes, and activities were appropriate.

Embark teachers' use of shared symbols and language was well-illustrated in one after-school professional development meeting when Embark's principal, Ms. Rivera, asked the assembled group: "How do you know the conditions of learning are supported?" The teachers answer synchronously, in raised voices: "Academic rigor! Accountable talk! Rubrics!" The principal nodded with apparent approval and responded, "See, these are all words you know!" While the exchange might be taken as a *pro forma* display of principal authority, it is less notable for what the principal communicated to teachers than for what teachers communicated to each other. Their coordinated use of specialized language, such as academic rigor, accountable talk, and rubrics, cued underlying meanings and values, and signaled understandings and expectations about shared approaches to instruction.

An observation at a similar training session revealed how teachers sometimes used subtle signals and body language to communicate their agreement or disagreement with colleagues. Because indications of agreement validated teachers' views, agreement was a form of social reward; indications of disagreement repudiating teachers' views were a form of social sanction. The use of social sanctions and rewards was poignantly illustrated by two male teachers and a female colleague seated at a round table near the back of the room during a professional development session where the principal played a video demonstrating how to differentiate instruction:

> Teacher 1 (aloud to colleagues seated nearby): You need to create 30 lesson plans for that. I don't see how he has the time! That takes time. That takes a lot of assessment.
>
> Teacher 2: (Nods vigorously in agreement).
>
> Teacher 3: (Shrugs and rolls her eyes). I know. They know. They know it's not realistic.
>
> Teacher 4: (leaning over from a nearby table, whispering) That kind of thing could never work here.

Rather than receiving specific social sanctions or rewards Embark teachers were sometimes influenced by the mere fact that a superior number of colleagues had certain attitudes and engaged in certain activities. For example, the importance of what a lot of other teachers were doing was expressed by Ms. McGintey, a seventh-grade teacher, who explained:

> You know, a school of this size and of this nature, you're always going to have some group of people resistant to change. When you hear from a lot of people that they don't agree with the change, that they're not going to do anything different, you ask yourself, "how much does it really matter if I make a change?"

Sometimes Embark teachers deliberately sought to influence their colleagues by explicitly instructing colleagues in the practices they felt colleagues should know. One teacher explained how he tried to use his past experiences to influence newer teachers who had been in the same teacher training program as him: "I'll say to them, 'Well, last year I tried this. I did this. Why don't you try this?'". Another teacher described intra-grade efforts to coordinate instruction this way:

> All the seventh graders are on the third floor, which is good because then teachers can talk to each other about what they are doing. If I'm trying out that new unit and it just doesn't seem to work, I can ask Ms. Peters what she is doing . . . But you know, when everyone you need to talk to is in the same grade it can be a bit tough in terms of looking for help from each other because of your schedules. They don't always coordinate. When you're teaching they're free; when you're free they're teaching. You kind of have to catch each other on the fly.

This teacher's description of attempts to coordinate instruction illustrates how influence and updating among teachers could result from teachers' mutual efforts to match their instructional approach. This matching process helped to generate shared instructional understandings. Note, however, that the above teacher's description of coordination also illustrates some of the limits of influence among Embark teachers. Specifically, different Embark teachers were more likely to be influenced by some teachers rather than others, namely the teachers with whom they interacted most frequently based on schedules or other factors. These patterns of interaction are important because, at Embark, they may have been the difference between socio-cognitive routines achieving a new, dynamic equilibrium given wide-spread use of the workshop model or the system returning to its steady-state, meaning the continued use of previous instructional practices. Ultimately, the latter proved true at Embark.

Four years after reform began, Embark teachers were unable to make sufficient changes in their instructional practices and the school was closed by the state. The state cited the school's stagnant achievement as the reason for its closure. When asked about the school's closure in a post-study, follow-up interview, one of Embark's former assistant principals who had gone on to work with a charter management organization, described the reasons for Embark's closure this way:

> How could I say this? Changing that school, really being able to change that school had to be something that you believed in full-fledge. It's not something that you could have picked up at 8:30 and dropped it off at 3 o'clock. See? It was always going to be a lot of work. Everybody had to be willing to change, everybody had to be on board to make that commitment. It couldn't just be some people. It had to be everybody getting better for the kids – better planning, better instruction. Everybody agreeing "it's not about you; it's for these kids." That just never happened. It's just something that everybody never came to understand.

## Discussion: Highly Differentiated Systems and Static Equilibria

### *The Consequences of Fragmented Social Structure*

One reason "everybody never came to understand" the need for new instructional standards at Embark was partly because everybody rarely interacted. With 68 teachers, Embark was a large middle school. Embark teachers taught in three different grades and nine subject specialties in classrooms on three floors of hallways that spanned almost a full city block. Teachers had two break periods during any one of seven 50-minute intervals from 8:30am to 3:30pm. As a result, teachers had "common preps" with some teachers, but little opportunity to interact with other teachers except before and after school.

Beyond the strictures of subject, classroom location, and schedules, which are typical structural features of schools, some particular features of Embark's culture further undermined teachers' opportunities to get to interact. For example, as part of the school's new safety protocol, administrators required that the school's entryway remain cleared of staff and students. Thus, in contrast to schools where teachers might congregate at the front entrance as they arrived at school, directives from Embark's security guard asked that teachers "please move on inside" or "please help us keep this area clear." These directives typically disrupted teachers' fledgling conversations and dispersed teachers to their respective classrooms, thus preempting potential opportunities for teachers to interact with colleagues they might not already know.

The patterns of teacher interactions at Embark resulted in teachers mainly interacting in identifiable 4–8 member communities based on geographic, temporal, and social proximity. As a result, Embark teachers also had more limited opportunities for collegial comparisons. Instead of making comparisons to a large number of teachers across the school, Embark teachers most frequently made comparisons to a small number of teachers within their communities. This also meant Embark teachers received less abundant peer feedback about instruction than they otherwise might have and experienced weaker influence to update their practice since feedback and pressure were coming from fewer sources.

This problem was mitigated only marginally by teachers who spanned boundaries between communities, such as an Embark teacher named Mr. Acevedo who shared membership in the community of new teachers who met most often in the teachers' lounge and the community of Spanish-speaking teachers, who tended to be longer-tenured and congregated in the school library. Such boundary spanners tend to be more peripheral members of communities and so have fewer influence opportunities to exert influence and may be weaker sources of influence given their status in each community (Bridwell-Mitchell, 2016; Lave & Wenger, 1991). The limited effects of boundary spanners in addition to other characteristics of Embark's social structure made the widely shared understandings necessary for reform especially difficult at Embark.

## *Sub-goal Differentiation and Integration in Complex Adaptive Systems*

The effects of social structure on Embark teachers' ability to generate widely shared understandings about reform is one illustration of the broader problem of coordinating action in highly differentiated systems, meaning systems in which functions are distributed across numerous subsystems (Ashmos & Huber, 1987). In organizational analyses, the relevant subsystems are often conceptualized as formal units, such as different departments or divisions (Scott & Davis, 2007). At Embark, the subsystems were *informal* units, meaning the numerous, small communities determined by teachers' informal patterns of interaction rather than strictly by their formally assigned roles and responsibilities in the schools. Despite being informal, these community subsystems were still responsible for certain organizational functions, in that community interactions determined teachers' instructional attitudes and activities, as described throughout the above case.

One characteristic of subsystems is that they have local sub-goals, which ideally align and support the overall system goal. Yet, it is possible for subsystems to converge on sub-goals that are local optima but that are suboptimal with respect to overall system goals (Amaral & Uzzi, 2007). This is to say that, in the case of Embark, instructional attitudes and activities in a given teacher community might be optimal or desirable for that community but might not serve the overall goals of the school. Even if the local goals of each subsystem do serve overall system goals, the absence of a mechanism for coordinating or integrating activities and outputs in each subsystem can still undermine overall system goals (Lawrence & Lorsch, 1967). Alternatively, conflicting sub-goals across different subsystems might cancel out potentially positive effects (Krackhardt, 1997).

What this meant for Embark and its response to external mandates for instructional reform is that even if instructional attitudes and activities among some teacher communities supported reform overall, reform could have been undermined for at least three possible reasons. First, the existence of numerous small communities but the absence of reliable mechanisms for coordinating their work meant teachers in different communities may have been unaware of reform-oriented work in other communities and so did not make the requisite adjustments. Second, teachers in different communities may have been aware of reform-oriented work of other communities. However, more frequent comparisons, feedback, and influence within their cohesive community might have suggested that adjustments in instructional attitudes, aims, and activities were unnecessary or inappropriate (Coleman, Katz, & Menzel, 1957). Finally, even if teachers in a given community were aware of and recognized the need for adjustments, it might have been difficult to determine exactly which adjustments to make, given different models across different communities. As a result, teachers might make no adjustment.

All three of the above scenarios would have the same effect at the overall system level – the reinforcement of the static equilibrium, which is to say that existing instructional attitudes, aims, and activities would remain unchanged. The important role that social structure played in system dynamics at Embark suggests the need to revise Model 1 for Socio-Cognitive Control Processes for System Responses to Exogenous Shocks. An additional statement is needed to explain the contingent expectations for adjustments in standards for work functions given the possibility of multiple, competing social standards across subgroups. Specifically, the argument – formalized using natural language first-order logic operator in the revised model below – is that if social standards across subgroups do not correspond and there is insufficient comparison or feedback across subgroups then no adjustment in standards via social influence results. The expectation is that this is the case both for the subgroup and the overall organization.

**MODEL 1.1.** Socio-Cognitive Control Processes for System Responses to Exogenous Shocks

---

Statement 1: GIVEN ANY external demand requiring new work functions (∀ED)

1.1 THERE EXISTS uncertainty about standards for work functions (∃UN)

Statement 2: GIVEN uncertainty actors employ social comparisons to determine standards (SC/UN)

Statement 3: Social comparison IMPLIES actors receive feedback GIVEN observed practices AND conversations about practice ($SC \rightarrow FB/OP \wedge CP$)

Statement 4: IF feedback DOES NOT CORRESPOND with the social standard THEN adjustments via social influence result ($FB \neq SS \supset AJ$)

4.1 IF feedback CORRESPONDS with the social standard THEN adjustments via social influence do NOT result ($FB = SS \supset \sim AJ$)

Statement 5: IF social standards across subgroups, $W_{ik\text{-}jk}$, DO NOT CORRESPOND AND THERE DOES NOT EXIST sufficient social comparison or feedback THEN adjustments via social influence do NOT result ($SS_{Wik} \neq SS_{Wjk} \wedge \sim \exists\, SC \vee FB \supset \sim AJ$)

## Conclusion

The premise of this chapter has been that the system perspective provides a useful lens for understanding and explaining school operations and outcomes, which might otherwise seem idiosyncratic or inexplicable. The intent is not to argue that the systems perspective should replace other lenses for analyzing organizational dynamics or educational phenomena. Indeed, in complex systems, many complex dynamics occur simultaneously. Thus, even in what appear to be alternative explanations, the basic principles of systems likely play a role and so should be considered as a way to further understand organizational dynamics and educational phenomena.

Indeed, the focal case in this chapter has been analyzed, in part, elsewhere through the lens of neo-institutional theory (Bridwell-Mitchell, 2016). The neo-institutional lens is especially applicable to the case, given the observed pressures from regulatory actors (i.e., the state department of education) and given that existing instructional practices at the school were long-held, deeply ingrained and value-laden despite being ineffective (DiMaggio & Powell, 1983; Meyer & Rowan, 1977). However, had the case findings initially been analyzed with the principles of system dynamics in mind, the important role of counter-balancing forces enabling or inhibiting teacher learning and institutional change may have been much more immediately clear. One conclusion to be drawn is that phenomena resulting from institutional processes are, at the same time, being driven by system dynamics. The same is likely true for other educational phenomena analyzed through other ontological lenses.

One additional benefit of the systems perspective is that its conceptual and analytical tools provide a way of structuring arguments and analyses. This is by virtue of providing a model or framework against which emerging arguments and findings might be compared. This also is by virtue of providing methods for formalizing arguments so that they are more concrete and more easily falsified (i.e., Maguire et al., 2006). One example of this kind of formalization is the first-order logic statements used in this chapter to model observed sociocognitive systems control processes. Given the potential utility and explanatory validity, the systems perspective may be an especially high-leverage conceptual and analytical tool for researchers and practitioners intending to better understand and improve school operations and outcomes.

## Notes

1 The methods from which this case was generated are detailed elsewhere. See Bridwell-Mitchell E. N. (2016). Collaborative institutional agency: How peer learning in communities of practice enables and inhibits micro-institutional change. *Organization Studies, 37*(2), 161–192.
2 All names are pseudonyms.

## References

Amaral, L. A. N., & Uzzi, B. (2007). Complex systems – a new paradigm for the integrative study of management, physical, and technical systems. *Management Science, 53*(7), 1033–1035.

Ashmos, D., & Huber, G. (1987). The systems paradigm in organizational theory: Correcting the record and suggesting the future. *Academy of Management Journal, 12*(4), 607–621.

Bidwell, C. E. (1965). The school as a formal organization. In J. March (Ed.), *Handbook of organizations* (pp. 972–1022). Chicago: Rand McNally.

Boulding, K. (1956). General systems theory – the skeleton of science. *Management Science, 2*(3), 197–208.

Bridwell-Mitchell, E. N. (2016). Collaborative institutional agency: How peer learning in communities of practice enables and inhibits micro-institutional change. *Organization Studies, 37*(2), 161–192.

Callon, M. (2009). Elaborating the notion of performativity. *Le libellio d'AEGIS, 5*(1), 18–29.

Coburn, C. E. (2001). Collective sensemaking about reading: How teachers mediate reading policy in their professional communities. *Education Evaluation and Policy Analysis, 23*(2), 145–170.

Coleman, J., Katz, E., & Menzel, H. (1957). The diffusion of innovation among physicians. *Sociometry, 20*(4), 253–270.

DiMaggio, P., & Powell, W. (1983). The iron cage revisited: Institutional isomorphism and collective rationality in organizational fields. *American Sociological Review, 48*, 147–160.

Dunn, W. N., & Ginsberg, A. (1986). A sociocognitive network approach to organizational analysis. *Human Relations, 39*(11), 955–975.

Fenwick, T., & Edwards, R. (2010). *Actor-network theory in education*. London: Routledge.

Festinger, L. (1954). A theory of social comparison processes. *Human Relations, 7*, 114–140.

Fitzpatrick, A. R. (1989). Social influences in standard setting: The effects of social interaction on group judgments. *Review of Educational Research, 59*(3), 315–328.

Ford, J., & Ford, L. (1995). The role of conversation in producing intentional change in organizational. *The Academy of Management Review, 20*(3), 541–570.

Forrester, J. W. (1968). *Principles of systems*. Cambridge, MA: MIT Press.

Frank, K. A., & Fahrbach, K. (1999). Organization culture as a complex system: Balance and information models of influence and selection. *Organization Science, 10*(3), 253–277.

Friedkin, N. E. (2001). Norm formation in social influence networks. *Social Networks, 23*, 167–189.

Hannan, M. T. (1998). Rethinking age dependence in organizational mortality: Logical formalizations. *American Journal of Sociology, 104*(1), 126–164.

Heise, D. R., & Durig, A. (2000). Qualitative models. In E. F. Borgatta & R. J. Montgomery (Eds.), *Encyclopedia of sociology* (Vol. *IV*, pp. 1582–1586). New York: Macmillan.

Hill, H. C. (2001). Policy is not enough: Language and the interpretation of state standards. *American Educational Research Journal, 38*(2), 289–318.

Holland, J. (1995). Can there be a unified theory of complex adaptive systems? In H. Morowitz & J. Singer (Eds.), *The mind, the brain and complex adaptive systems* (Vol. *XXII*, pp. 45–50). Reading, MA: Addison-Wesley Publishing Company.

Kast, F. E., & Rosenzweig, J. E. (1972). General systems theory: Applications for organizations and management. *Academy of Management Journal, 15*(4), 447–465.

Katz, D., & Kahn, R. L. (1978). *The social psychology of organizations*. New York: Wiley.

Katz, E., & Lazarsfeld, P. (1966). *Personal influence*. New York: The Free Press.

Krackhardt, D. (1997). Organizational viscosity and the diffusion of controversial innovations. *Journal of Mathematical Sociology, 22*(2), 177–199.

Latour, B. (1991). Technology is society made durable. In J. Law (Ed.), *A sociology of monsters: Chapters on power, technology and domination* (pp. 103–131). New York: Routledge.

Lave, J., & Wenger, E. (1991). *Situated learning: Legitimate peripheral participation*. Cambridge: Cambridge University Press.

Lawrence, P., & Lorsch, J. (1967). Differentiation and integration in complex organizations. *Administrative Science Quarterly*, 1–47.

Maguire, S., McKelvey, B., Mirabeau, L., & Oztas, N. (2006). Complexity science and organization studies. In S. Clegg & C. Hardy (Eds.), *Sage handbook of organizations studies*. London: Sage Publications.

Merton, R., & Rossi, A. K. (1957). Contributions to the theory of reference group behavior. In R. Merton (Ed.), *Social theory and social structure*. New York: Free Press.

Meyer, J., & Rowan, B. (1977). Institutionalized organizations: Formal structure as myth and ceremonies. *American Journal of Sociology, 83*(2), 340–362.

Miller, J. G. (1972). Living systems: The organization. *Behavioral Science, 17*(1), 2–182.

Moore, J. C. (1968). Status and influence in small group interactions. *Sociometry, 31*(1), 47–63.

Morel, B., & Ramanujam, R. (1999). Through the looking glass of complexity: The dynamics of organizations as adaptive and evolving systems. *Organization Science, 10*(3), 278–293.

Payne, C. M. (2008). *So much reform, so little change: The persistence of failure in urban schools*. Boston, MA: Harvard Education Press.

Pollock, J. L. (2002). The logical foundations of means-end reasoning. In R. Elio (Ed.), *Common sense, reasoning, and rationality* (pp. 60–77). New York: Oxford University Press.

Rivkin, J. W., & Siggelkow, N. (2007). Patterned interaction in complex systems. *Management Science, 53*(7), 1068–1085.
Scott, R. (1998). Chapter 4: Organizations as open systems, *Organizations: Rational, natural and open systems*. Upper Saddle River, New Jersey: Prentice Hall.
Scott, W. R., & Davis, G. F. (2007). Chapter 6: Technology and structure, *Organizations and organizing: Rational, natural, and open systems perspectives* (pp. 124–150). Upper Saddle River, NJ: Prentice Hall.
Sherif, M. (1966). *The psychology of social norms*. New York: Harper and Row.
Spillane, J. P., Reiser, B. J., & Reimer, T. (2002). Policy implementation and cognition: Reframing and refocusing implementation research. *Review of Educational Research, 72*(3), 387–431.
Tarski, A. (1946). *Introduction to logic*. New York: Oxford University Press.
Von Bertalanfy, L. (1972). The history and status of general systems theory. *Academy of Management Journal, 15*(4), 407–426.

# 2

# SYSTEMS THINKING TO DRIVE SCHOOL TURNAROUND

*Sharon F. Rallis and Rachael B. Lawrence*

Many U.S. public schools in high-needs urban areas operate in and with failing systems. Success in teaching and learning in such schools is challenging and often impossible (Anderson & Stillman, 2013). In recent decades, considerable U.S. federal dollars have been directed toward reforming these schools. Few have been successful in either the short or long term (Dragoset et al., 2017). We suggest that too many reform efforts lack attention to systems and the systemic thinking required to bring about sustainable, enduring change. Riverton, an urban elementary school located in Portland, ME that in 2009 was designated as a *persistently failing school* by the U.S. Department of Education but demonstrated a genuine *turnaround* by 2013, illustrates a case of holistic change guided by systems thinking (Lawrence, 2016). We hope presenting our experience studying this school can provide insights regarding the courage, risks, and actions that schools must take to address the interconnectedness of human interaction to enact system change.

School turnaround literature indicates that the right leader is essential to improvement (Dodman, 2014; Duke, 2015; May & Sanders, 2013; Reyes & Garcia, 2014). A single-loop thinking theory of action might lead to the conclusion that strengthening the leader may be the best approach to school improvement. However, double-loop thinking suggests that to change the school a cadre or league of leaders engaging together in inquiry and action is needed to create and nurture the change (Militello, Rallis, & Goldring, 2009; Rallis & Militello, 2010). In our case example, the Chief Academic Officer (CAO), the principal, the literacy coach, union leaders, and teachers worked together to question and change teaching practice in their schools.

An operating theory of action in Riverton was: "if teachers learn, students will learn." Originally this theory functioned with single-loop thinking that encourages individual teacher professional development to improve instruction (Rallis, Churchill, Lawrence, & Darling, 2011). As systemic thinking took hold, double-loop learning dominated with recognition that teachers need to learn together around common instructional themes. Using the Riverton case, we describe how systems thinking by key actors in the school and district led to dramatic changes that facilitated a holistic culture of teaching and learning in a U.S. elementary school populated by many high-needs learners.

We define systems as consisting of elements – the parts that make up the whole – that are linked together by processes and interrelationships; these parts are bounded, limiting what falls inside the system (Williams & Hummelbrunner, 2011). *Systems thinking* is a way to understand systems holistically, focusing on how the parts work together over time to function successfully (Shaked & Schechter, 2017). We draw on institutional theory that describes an organization's processes and relationships as regulatory, normative, or cultural-cognitive (Scott, 2005, 2008). As well, our analysis is informed by the classic systems thinking work by Argyris (1976), who posits that focus on a single variable in problem solving or learning leads to flawed assumptions and double-loop learning asks us to question our assumptions from a systemic point of view. In this chapter, we tell the story of the change at Riverton, first describing the context of the school and our research, our conceptual framework for systemic change, and our analysis of what happened.

## Context: The Story of the School and our Studies

We are researchers from the University of Massachusetts Amherst who first came to Portland Public Schools (PPS) in 2011 to examine the effects of a knowledge and skills based salary schedule proposed by the union, approved by the school board, and implemented in 2009. Portland's Professional Learning Based Salary Schedule (PLBSS) was grounded in the theory of action that teacher involvement in targeted quality professional learning could positively affect student learning: put simply, *if teachers learn, students learn.* The program argued that traditional professional development (PD) failed to produce authentic teacher learning because teachers had little opportunity to choose PD that was directly related to their classroom instruction. PLBSS gave teachers great control over their PD options and rewarded them for these choices. Under this system, teachers could participate in book groups, study sessions, or take courses on what they perceived as being their most urgent instructional needs, and they would advance on the salary schedule for their participation in instructionally relevant PD. Moreover, the union and school board anticipated that this salary system would improve both recruitment and retention of quality teachers. Our original study revealed that district hopes were met: salaries increased; teachers stayed in the district; and any vacancies within the district were readily filled with strong candidates. Surveys and interviews indicated that teachers were selecting PD according to what they perceived were their students' needs. Records revealed high rates of participation and variation in the kinds of PD experiences offered and chosen. Often groups of teachers chose or designed a course together. Both leadership and teachers reported that participation in the PD improved instruction and provided compelling anecdotal evidence for improved student learning. We chose not to look at achievement scores since we could not, with validity, attribute any increases or decreases to the PLBSS (Rallis et al., 2011). Intrigued by what we heard, we returned to explore if we could make connections between participation in PLBSS and student achievement.

To design the second study, we worked with district and union leaders to select four elementary schools in which we carefully examined the teachers' professional development participation records to compare with student testing outcomes. We coded the PD database for teachers at the four sites, and then looked at the rate of test score change at each school. In doing so, we came across what we thought was an anomaly. One of the four schools had almost zero participation in PD recorded in the past three years. And yet, this school's test scores appeared to be rising more rapidly than the others in a statistically significant way (Rallis, Keller, Lawrence, & Soto, unpublished study). Other schools in the study had robust participation in this record system, and yet their scores were not remarkable. We thought perhaps something was amiss – so we asked why Riverton seemed to have no participation in the PLBSS system.

"Go in and see it!" was the response from union leaders. "They have been working so hard, and there's a reason you can't see it in our database. We're working on it for them, though" (interview, 2013).

Something different than we expected, given the demographics, the location of the school, and the activity recorded in the professional learning database was happening, and we had to go see what it was. As a seasoned professor and researcher of education reform and a former teacher who worked in high-needs schools, what we found at Riverton was truly unexpected. We witnessed a case of people thinking and working together systemically to enact whole school positive change.

## The Case

Riverton School is located in Maine's largest city. In 2013, 73.2 percent of students qualified for free and reduced lunch; 37.6 percent of the students were English Language Learners; 49 percent of the students spoke another language at home (Maine Department of Education Data Warehouse, 2017). All told, more than 23 separate languages and dialects were spoken in the homes of Riverton students. In recent years, Portland has become home to many refugees – families fleeing conflict zones and political oppression. As a result, students and their families came from places like Russia, Somalia, Vietnam, Afghanistan, and other nations (interviews, 2013).

Even before the recent wave of immigration to Portland changing the student population to high ELL needs, Riverton had long been considered a less-than-adequate school. Looking back on the school, one leader remarked, "For a decade before, you could have run any 400 kids through that school, and the results would have been the same . . . the adults [would have said] 'those kids can't learn'" (interview, 2013). Teachers reported that, in the past, they had blamed parents for not sending their kids to school ready to learn (interviews, 2013). And, the building was old, poorly maintained. "Riverton is dark, like in a deep hole," recalled a district leader (interview, 2013). Teachers reported that they felt isolated in the building (interviews, 2013; 2014). Neighborhood parents also felt something was wrong. They made this clear in their actions by complaining on the district polls (interview, 2014) or by sending their kids to parochial options when they could afford to do so (parent conversation, 2014). This school had desperately needed to change the way things were done.

The state of Maine recognized that problems existed and identified Riverton as a *persistently failing school* in 2009. With that label, Riverton became eligible for a federally funded School Improvement Grant (SIG), part of the American Reinvestment and Recovery Act passed earlier that year. Portland Public School District applied for the SIG and were awarded a substantial grant ($3,386,154.00) to support change in the school. Professionals across the district pulled together to plan the change project; central office administrators collaborated with school officials, the union, and teachers to develop a plan.

Entering the school for the first time in 2013, we never saw the "persistently failing" Riverton in a "dark hole." We never saw the discouraged teachers, the students with unmet needs, or the run-down school. In fact, the only evidence we have of Riverton's past is the testimony of those who had lived the change. The meteoric rise in test scores also demonstrated that something different was happening in Riverton than in the past. The school we entered was bright, organized, cheerful, welcoming. As we walked in the front door we were greeted by a sign with the word *welcome* translated into the 23 languages spoken. People smiled as we met them. Student work was proudly displayed in the halls, and as we walked around, we could see that these selections were purposeful and interconnected. When we interviewed teachers, we heard:

> I know how to teach my kids.
>
> The kids love coming to school now.
>
> My students are getting to a deeper level of understanding.
>
> *(Interviews, 2013; 2014)*

Teachers expressed improved efficacy, demonstrating confidence in their ability to teach to reading and writing. Students seemed happy and content, even proud of their work. The students' standardized test scores increased across all tested areas. The most dramatic change was in reading and writing, but students also improved performance in mathematics. The culture of teaching and learning at Riverton had entirely transformed from that described by school officials as they looked back at the time before the turnaround efforts (Lawrence, 2016).

Riverton is a Turnaround School that actually turned-around. What happened?

## Conceptual Framework: Systemic Thinking about Change

We argue that the change in Riverton School resulted from people questioning their assumptions and learning to work together in new ways. Their thinking became systemic, combining inquiry with action (see Militello et al., 2009). Any system consists of linked elements interacting within the whole. Multiple systems were involved in the turnaround since Riverton was itself nested within other systems: district, state, and federal education. While many of the elements in the larger systems remained the same, people directly involved with Riverton engaged in inquiry that led to rethinking their practices: they *observed*; they *identified* and *focused* on problems; they consciously *examined* and *re-conceptualized* these problems; and they *experimented* and *took action* (Schön, 1983). They moved from seeing only the individual teachers, students, classrooms, families toward making sense of how these parts interacted with each other and were related to the central element (represented by the union and Chief Academic Officer). This recognition that every action (or inaction) by one part affected other parts, which then affected others and the whole, shaped their operating perspectives. Their systemic thinking altered roles and the ways people interacted – they were willing to take risks to act in new ways.

We, the research team, found the best way to understand what happened at Riverton was to use systemic thinking ourselves; thus our research methods were reflective, collaborative, assuming evolution, iterative. Our ethnographic approach was itself holistic, balancing between the elements and the whole. "Systemic thinking is about making sense of the world rather than merely describing it. It is fundamentally a process that organizes the messiness of the real world into concepts and components that allow us to understand things a bit better" (Williams & Hummelbrunner, 2011). To start organizing the messiness of the change, we asked: What relevant elements were operating? How were these elements connected to each other and to a center? Thus, our first organizing concept is *boundary* (see Williams & Hummelbrunner, 2011, for expanded discussion of these concepts). Because we acknowledge that any system is a mental construct, we wanted ours to be manageable. We considered how we would determine which elements would fall inside our system of focus – and which would not. We also considered whose voices were important to define where boundaries would fall, so we sought input from leaders across the systems. The CAO, the union professional development coordinator, and the union president also wanted to understand the change, so they agreed to work closely with us. Together we decided to draw boundary around the school itself. While acknowledging that the school was embedded in multiple external interrelationships, we felt that the place to start was with what happened within the perimeters of the school property. Thus, we attended only to elements within the school; decisions and actions of teachers, students, administrators, CAO, union officials, and parents were only considered when they were either present inside or directed to the school.

Defining or framing our system with the limit of the school is not a neutral act. Although we recognized that external relationships did influence the situation, we were establishing a particular perspective, assigning importance and value to internal events, actions, roles, and relationships. How did our boundary align with boundaries others have drawn? Would our analyses and syntheses be acceptable to inform practice and to serve the systems in which Riverton is embedded? Was our boundary fair? That is, were we marginalizing, silencing, or excluding some while privileging others? (Rossman & Rallis, 2010). What unintended consequences might our framing have? Would it shape our expectations and, as a result, our findings? Designing the study closely with the leaders addressed these questions. We all agreed at the outset to make our reasoning and decisions transparent and to remain open to alternate views or interpretations.

The next organizing concept that logically follows boundaries is *perspectives*. Perspective or orientation shapes or frames *how* we see or interpret a situation. Perspectives are deeply rooted in sets of assumptions held about reality (does reality exist independent of my perceptions? How have my experiences shaped my perceptions of reality?) and how reality is discovered (what do I accept as evidence? How do I learn?) (see Rallis & Rossman, 2012; Rossman & Rallis, 2017). To think systemically, individuals have to be aware that any situation can be seen from more than one perspective – and that different people use different lenses to determine how their research develops: what data to collect, what is considered legitimate evidence, how to make interpretations.

> Perspectives importantly draw the focus away from the system or situation as it supposedly exists in "real life" and allow us to consider alternatives: what it might be like, could be like, or even should be like . . . The similarities and differences between perceptions of *what is, what people think it is*, and *what people think it could be* create puzzles and contradictions that achieve deeper learning.
>
> *(Williams & Hummelbrunner, 2011, p. 21)*

Our research team had both a qualitative expert and a quantitative expert, so internally we represented more than one perspective. We also brought varied experiences regarding schooling (teacher, principal, school board member, testing expert). Similarly, the people at Riverton bought various worldviews that may have differed from those of the district leaders. Our dialogues helped us make sense more holistically. For example, union and central office leaders were driven at the outset by socio-political forces that demanded proof of effectiveness. They wanted evidence that would link specific decisions (e.g., embedding an instructional coach; school driven PD) with the increased scores. Our researcher lens fought back, questioning the legitimacy of attributing any cause-effect linkages. As we talked together, we asked how we could document actions and interactions and then make sense of them. We also sought ways to better inform the attribution questions. What perspectives did the participants themselves hold? How did they see the changes in Riverton? What influenced these perspectives? We planned the research together and regularly shared and discussed our findings.

Identifying perspectives provides insight into another organizing concept: *interrelationships*. System elements are linked; processes and relationships hold them together within an accepted boundary defined by a central concept (*student learning* in the Riverton case). How are the elements connected? Where are connections strong or weak? Multiple theorists suggest that simple, complicated, and complex decision-making domains exist within systems theory (see Gell-Mann, 1995; Kurtz & Snowden, 2003; Patton, 2011; Rogers, 2008). We propose, drawing on common descriptions of chemical bonds and electronic network systems, that these categories can describe interrelationships. The connections could be *simple* with weak links between elements but strong linkages to the center; they could be *complicated* by strong links between both other elements and the center; or, they could be *complex*, strongly connected to each other but with no center. Interrelationships that exist within a system depend on context and the needs of the elements in that context. Interrelationships change as the system changes. One type is not inherently better than another; rather, different contexts call for different interrelationships.

Various processes guide the interrelationships and connect the elements. Are they regulatory, normative, or cultural-cognitive (taken-for-granted) (Scott, 2005, 2008)? What patterns can we see? For example, are they dynamic, linear, or nonlinear? What feedback loops exist? How sensitive to the context are the relationships? What consequences do the connections have? Argyris (1993) posits that learning occurs when a mismatch exists between a goal, intent or value and the consequences of action. What he calls single-loop learning results in modifying only the action; whereas double-loop learning calls for a re-examination of the underlying assumptions that drive the actions, what Argyris labels governing variables.

Exploring these questions and making sense of the *interrelationships* in terms of the changes at Riverton is an interpretive act; it requires that we look at the human actions with regard to specific situations: what do we think people are doing and why? What happens around the actions and how do we think these are related? We believe that human actions are constructions; people act in response to how they have defined others' actions, assuming the other's intent.

People interpret each other's actions as the means of determining how to act toward one another (see Blumer, 1969). Sometimes participants in a situation share interpretations of actions; often they do not. Meanings become attached to certain actions, which can become symbolic, serving regulatory, normative, and cultural functions.

In summary, we use boundary, perspective, and interrelationship to guide our analysis and interpretations and to understand how a league of leaders engaging in double-loop learning, through an inquiry-action cycle, helped turn Riverton around. As researchers, our making sense of a situation is actually recursive interpretation: making sense of the participants making sense of each other. Our data come from record reviews, test score analyses, and two years of visits to Riverton for ethnographic interviews and observations. We divide the change process into three phases. Elements presented throughout the process included the following: teachers, principal, literacy coach, union coordinator and president, and deputy superintendent.

## Riverton Turns Around: Making Sense of What Happened

### *Phase I: A Simple System*

When Portland Public Schools received the School Improvement Grant, federal policy dictated that schools should use one of four models to produce change as a condition of the grant: Turnaround, Transformation, Restart, or School Closure. In the first year of funding, the school elected to use the Transformation Model, which required schools to work towards *improving teacher and leader effectiveness, extending the school day, fostering community orientation*, and *providing operational flexibility* (Terry, 2010). To improve effectiveness, the Chief Academic Officer (CAO) and the school's full-time literacy coach, Tracey, made the decision to adopt a common curriculum. They selected a well-known reading and writing program, *Units of Study*, commonly called the *Writers and Readers Workshop*. Further, the CAO and the Portland Education Association (PEA – the union) supported several actions: increasing the part-time literacy coach to full time to facilitate curriculum implementation and engaging faculty in expert level professional development at Teachers College in New York City. The school day was extended. To foster community engagement, resources were directed to create the School and Community Center in the building. Operational flexibility was minimally addressed by embedding the reading coach and by adjusting bus schedule to allow the extended day. As well, the district directed resources to repaint the school, changing its appearance to a bright, open space, very different from the dark walls that were an artifact of the original 1970s building. The repainting, the PD experience at Teachers College, and the School and Community Center became visible, even tangible symbols that change was occurring.

During this phase, many elements inside Riverton had not changed; teachers and the principal were those already in place. At the same time, the boundary extended outside the school building to include the PD sessions at Teachers College, which were reported as critical to teachers' learning to instruct Riverton's students. The teachers reported it was "inspirational" to go into schools in New York City that had implemented the curriculum successfully and see what they'd learned in action (interviews, 2013). The boundary also included union leadership and CAO, who became the central elements, driving the change. We classified the system as simple; links between the individuals, teacher groupings, the principal, coach, and others in the building were weak. These elements appeared to be more strongly connected to the central elements (union and CAO). Interrelationships appeared to be defined primarily by regulatory processes, that is, adhering to the guidelines or rules of the organization or leadership. Teachers were seen as resistant to getting help to change their practice. For example, they complained to the CAO that the professional development provider (a recognized expert in the selected curriculum from out-of-state) was "condescending" and "out of touch" with the needs of Riverton (interview, 2013). While the Riverton faculty enjoyed escaping the boundary of their school for a while, they did not seem to appreciate an outside "expert" entering their boundary in the same way.

We consider that a simple interrelationship system, with the focus on hierarchical elements and implementation of centrally decided strategies, fit the existing school context and thus was likely what was needed to initiate the change process. In this initial phase, actors within the school had not yet had the opportunity to really engage in inquiry-action processes, and the league of leaders within the school had not yet formed. Thus, at the end of the first year, the CAO felt the potential transformation through this model was yet unrealized. "Transformation of this school wasn't going to happen. We had to turn it around" (interview, 2013). He chose to have Riverton adopt the Turnaround Model.

### *Phase II: A Complicated System*

The Turnaround Model required change in leadership and faculty in addition to the elements described in the Transformation Model. As part of the Turnaround plan, a new principal, Jeannie, was hired. Jeannie brought extensive experience in leading literacy and language acquisition in a large, metropolitan school district. Having grown up in the district, she knew Portland Public Schools well. Jeannie expressed that part of her role in the school was as "a cheerleader" (interview, 2013). She brought a new style of leadership to Riverton. She recognized the needs of her faculty to feel supported in the change, bringing food and drink to professional development meetings. She participated in PD alongside her faculty, recognizing the presence of other leadership forces in the school. As principal, she seeded the inquiry process in the school by encouraging teachers to observe and consider what they were seeing in their own and others' practice.

The new model also meant that half of the teaching force needed to be replaced. Working with the union, many teachers who had worked in Riverton moved to positions elsewhere in the district; many had been "counseled out" to find positions more in line with their philosophies and skills. New teachers arrived to fill their positions. The teachers who remained expressed their willingness to take necessary risks to learn new processes. The CAO addressed the faculty resistance to the external expert professional development provider by hiring a local curriculum expert, Kelli, to work directly with the coach (Tracey) and teachers in the school.

The school embraced an embedded professional development system. In this setting, embedded meant a holistic approach with groups of teachers engaged in professional learning largely during the school day and in the school. What was learned in professional development was then further supported by Tracey who regularly scheduled opportunities for observations and feedback in the classrooms with the teachers. Sometimes the faculty workshopped teaching scenarios outside of the classroom, role-playing as students. The professional development specialist, Kelli, came into the school on a regular basis to observe and instruct, leading small-group professional learning sessions during the school day. This style of professional learning allowed teachers to observe, to identify and focus on an area of practice that needed to be reconceptualized, and to experiment with changing their instruction. Working together, they could troubleshoot and adjust practice, supporting a holistic inquiry-action cycle in the school.

Some structural changes were made to support the embedded PD. The principal changed the master schedule to ensure all teachers at each grade level had consistent common planning time during which they used their inquiry cycle. To further support teacher collaborative learning and engagement, three full-time floating substitute teachers (subs) were hired. The subs were trained in the curriculum and were familiar to all of the students in the school, due to their full-time status. To this end, they were able to provide consistency in curriculum delivery and supported teacher interactions and their developing interrelationships.

Prior to the turnaround, teachers and leaders noted that no shared strategies or curriculum for teaching writing existed at Riverton. With the implementation of the common curriculum, teachers developed sets of collective practices as well as a shared language of instruction. This provided students with consistency in reading and writing instruction that had not been available in the school before the turnaround. Also, the common language and strategies enabled the teachers to communicate with each other more effectively – revealing stronger bonds between grade level teachers. Getting comfortable with the observations and feedback of others in the classroom setting allowed teachers to question and revise assumptions underlying their practices, thus incorporating double-loop learning. The most obvious example is the revised assumptions about what it takes to teach the Riverton students. Additionally, the principal implemented a cross grade-level literacy team. This team had regular meetings with the literacy coach (Tracey), the PD provider (Kelli) and principal (Jeannie), serving as a new leadership structure in the school, acknowledging the developing league of leaders in Riverton. These structural adjustments supported holistic change.

Another key difference in the new model requires some financial incentive to be provided to faculty who remain through the turnaround. At first, how Riverton teachers could receive salary incentive for persistence in the turnaround was unclear because the contract did not allow PD credits (PLBSS) to be applied to "in school" training. The leaders recognized that the embedded PD happening at Riverton was above and beyond typical "in school" PD – and sometimes called for long hours before and after the school day as well. The union and district recognized that this PD deserved credit, so they worked together to modify the contract so that the Riverton turnaround teachers could receive a salary lane-change worth of credit for persisting through the three-year turnaround. This adaptation of the contract is evidence of systemic thinking; central elements were willing to waive control to support needed holistic interconnections between the teachers.

The collective inquiry, the common curriculum and shared language, the embedded PD that involved groups learning together – all these normative processes strengthened links between elements in the building (teachers, Jeannie, Kelli, and Tracey) while maintaining the relationship with the center (CAO and union leadership). All three levels of leadership (district, union, school) remarked on their shared perspectives, investment, and interests. The CAO said: "When we need to work something out with the union – Sue is right here!" For example, when the central office feared that through PLBSS, teachers would not choose courses that would address core needs of the district, these leaders agreed to suggest amending the contract to require that selections fit five core areas; this modification was adopted and implemented. Such a *complicated* system operates with the strong interrelationships required for the safe environment that allowed double-loop learning to bring about the internal holistic changes occurring in Riverton.

### *Phase III: A Complex System*

The third phase of the Riverton's Turnaround happened after the grant period expired and the money was fully expended. We first entered the school in 2013, at the very end of the SIG-funded period. We continued to visit the school in the first year post-funding. We were curious: would the changes that occurred during the Turnaround be sustainable once the funding was gone? According to the CAO, the Turnaround had worked because, "[The teachers] worked through the culture of being blamed towards a culture of *showing the world*. [The district office] took away perceived obstacles to teacher learning. And then, the teachers had a quick win with student learning" (interview, 2013, italics added to reflect change in tone).

During our visits, we saw clear evidence that the *collaborative inquiry-learning-action* culture that took root during the Turnaround period flourished in the year after. "Teaching is a public activity at Riverton," reflected two administrators (interviews 2013, 2014). Rather than closing doors and teaching on their own, the doors remained open to visitors. These visitors included other teachers, the literacy coach, the principal, and students from local universities coming in to observe. We observed teachers asking each other questions in the hall, tracking down the coach to ask for feedback on an issue or idea, and reaching out to the principal with questions.

Embedded professional development sessions continued after the SIG expired, but the relationship with the professional development provider changed. During the Turnaround, most sessions had been directed by Kelli (PD provider) and Tracey (coach). They jointly determined the direction needed for each group of teachers. However, in the year after, Kelli expressed that she was so confident in the teachers' abilities to collaborate effectively that she was now "turning the control over to them. I come in and ask, 'how do you want to use this time?' And – it's often right in line with what I would hope they want to work on" (conversation during PD session, 2014). Teachers, in taking agency for their own inquiry processes, joined the league of leadership.

Perspectives had shifted. In the years before the turnaround, people working within and around the school expressed the idea that the teachers at Riverton were "good teachers," but the students were not sent to school prepared to learn. Following the turnaround, Riverton teachers and leaders recognized that while the teachers might have been knowledgeable, that they needed to learn different skills in order to meet the needs of the students. One leader said, "They did not believe those kids could learn. And by the end, everybody had seen their kids' writing – more than ever before, and coming to school wanting to write" (interview, 2013). He continued, "Changing the adult viewpoint – until teachers experienced that their PD learnings could make a difference for kids, they didn't truly understand that the kids could learn." The teachers learned what students in Riverton needed, and they learned how to meet these needs – and they could rely on each other as resources for helping to meet the needs.

Links between in-school elements continued to be strong as the connections to the central elements of the union and CAO were weakening, indicating a more *complex* system. The collaborative inquiry-based interactions in Riverton became *cognitive-cultural*, that is, "the way we do things around here" (Deal & Kennedy, 1982, p. 4). People learned from and with each other; actions resulted from multiple minds leading with shared focus and goals. We suggest that positive changes in such complex systems with strong internal linkages are potentially more sustainable because work is owned by the internal elements working together, supporting each other as needed in a complex whole. Reliance on the external elements is reduced.

## Conclusion

What we saw at Riverton was a school where *systems thinking* was the taken-for-granted way of operating across multiple elements. That is, people involved with the school focused on making the many and various parts work together over time (Shaked & Schechter, 2017); collaborative inquiry and action became ongoing standard processes. Groups questioned assumptions about relationships and perspectives to inform decisions that guided actions, thus allowing change to occur naturally and holistically. People recognized connections between problems and solutions; events were not seen as isolated.

If a child struggled with a particular learning, the issue was addressed systemically. Teachers came together with the principal, the coach, and the PD coordinator as needed, creating safe zones to identify, frame, and accept responsibility for problems. A problem belonged to all of them – not just to the child or an individual teacher. As well, the solution lay among all – no longer did teachers look to experts or specialists or the principal to fix a situation. They gathered data, made plans, took action, and evaluated the effects. Together, they reflected, revisited underlying assumptions, and reconceptualized their problems of practice. Their double loop learning meant that they chose not to do more of the same; instead they took risks grounded in their inquiry to fundamentally change their practices. In short, systemic thinking worked in Riverton because leadership was exercised through all the elements, at all levels. A *league of leaders* engaged together in inquiry and action taking risks to question assumptions with the double-loop learning needed to create and nurture positive and sustainable change (Rallis & Militello, 2010).

This turnaround did not happen overnight; indeed, we see the turnaround as the outcome of Riverton's journey through different systems configurations. When the change effort began, Riverton operated, as do many schools, with a simple system: teachers within or across grade levels seldom interacted professionally; teacher learning occurred individually, usually through university courses; rules, regulations, and decisions came from the central office. As the changes took hold, the system became complicated: teachers sought to align their work across and even between grades; teacher learning was embedded and collective; central office administrators and union officials supported teachers' efforts to do things differently, and teachers relied on this support. Eventually, Riverton operated as the complex system we have described; support and decision-making resided internally with little reliance on central leadership. For Riverton, this systems progression worked to facilitate and sustain change.

Lest we leave the impression that teaching and learning at Riverton remained nearly idyllic forever, we return to our systemic thinking as researchers and offer several concluding points and raise challenging questions:

- "Schools are almost always aboil with some kind of 'change'" (Elmore, 2000). Riverton's journey will continue. Its many elements will evolve and more will be added; needs and resources will increase and decrease; external factors and forces will vary. Thus, relationships, perspectives, and boundaries will shift. Will Riverton's improvement process continue to be deliberate with clear and shared instructional goals? In what ways will the system alter? Will a league of leaders continue? And how will these leaders make sense of the changes? Will their systemic thinking endure?
- Riverton School is, of course, still part of other and larger systems. Over time, how do various external systems affect and shape Riverton? How will boundaries be redrawn or permeated? Can Riverton remain buffered from external forces? Can their way of doing things withstand the pressures from external institutionalized expectations of what makes a school legitimate (Meyer & Rowan, 1977) – especially if standardized achievement scores do not continue to rise?
- Similarly Riverton's turnaround and systems thinking has potential to affect other systems. In what ways has the Portland Public School system – and the Maine Department of Education – learned from and used the improvements in Riverton? What policies can support in other schools the kind of systems thinking and systems change that Riverton experienced?

Still, Riverton Elementary School illustrates a successful systems thinking approach to change. Their theory of action in practice became: *If all adults in Riverton learn together, we can teach our students, and they will learn.* Yet, we wonder – was the move to their holistic understanding of their work intentional and purposeful? Or was the systems thinking generated by the unique evolving context – the particular way elements came together, boundaries formed and re-formed, and perspectives meshed? We suspect a *chicken–egg* situation occurred: a variety of forces created a safe space for systems thinking to flourish and thus to recognize and act on potential interrelationships between elements and perspectives. Riverton's example may not tell us exactly *how* to turn every school around, but it does show us how systems thinking can support decisions for change that *can* turn a school around.

## References

Anderson, L. M., & Stillman, J. A. (2013). Student teaching's contribution to preservice teacher development: A review of research focused on the preparation of teachers for urban and high-needs contexts. *Review of Educational Research, 83*(1), 3–69.

Argyris, C. (1976). Single-loop and double-loop models in research on decision making. *Administrative Science Quarterly, 21*(3), 363–375.

Blumer, H. (1969). *Symbolic interactionism*. Englewood Cliffs, NJ: Prentice-Hall.

Deal, T. E. & Kennedy, A. A. (1982). *Corporate cultures: The rites and rituals of corporate life*. Reading, MA: Addison-Wesley Publishing Company.

Dodman, S. (2014). A vivid illustration of leadership: Principal's actions propel struggling school's turnaround. *JSD: The Learning Forward Journal, 35*(1), 56–62.

Dragoset, L., Thomas, J., Hermann, M., Deke, J., James-Burdumy, S., Graczewski, C., Boyle, A., Upton, R., Tannenbaum, C., Giffin, J., & Wel, T. E. (2017). School improvement grant: Implementation and effectiveness. IES – U. S. Department of Education. Retrieved from https://ies.ed.gov/ncee/pubs/20174013/pdf/20174013.pdf

Duke, D. L. (2015). *Leadership for low-performing schools: A step-by-step guide to the school turnaround process*. Lantham, MD: Rowman & Littlefield.

Elmore, R. F. (2000). *Building a new structure for school leadership*. Washington, DC: The Albert Shanker Institute.

Gell-Mann, M. (1995). Complex adaptive systems. In *Santa Fe Institute studies in the sciences of complexity – Proceedings*, 22. Boston: Addison-Wesley.

Kurtz, C. F., & Snowden, D. J. (2003). The new dynamics of strategy: Sense-making in a complex and complicated world. *IBM Journal of Research and Development, 42*(3), 462–483.

Lawrence, R. B. (2016). *Turning around the culture of teaching and learning: A turnaround school success story (doctoral dissertation)*. Amherst, MA: University of Massachusetts Amherst. Retrieved from http://scholarworks.umass.edu/dissertations_2/698/

Maine Department of Education Warehouse (2017). Available at https://www.maine.gov/doe/dataresources/warehouse.html

May, J. J., & Sanders, E. T. W. (2013). Beyond standardized test scores: An examination of leadership and climate as leading indicators of future success in the transformation of turnaround schools. *Journal of Urban Learning, Teaching, and Research, 9*, 42–54.

Meyer, J. W., & Rowan, B. (1977). Institutionalized organizations: Formal structure as myth and ceremony. *American Journal of Sociology, 83*(2), 340–363.

Militello, M., Rallis, S. F., & Goldring, E. (2009). *Leading with inquiry and action: How principals improve teaching and learning*. Thousand Oaks, CA: Corwin.

Patton, M. Q. (2011). *Developmental evaluation: Applying complexity concepts to enhance innovation and use*. New York: Guilford.

Rallis, S. F., & Militello, M. (2010). The league of leadership. *Principal Leadership, 10*(8), 30–34.

Rallis, S. F., & Rossman, G. B. (2012). *The research journey: Introduction to inquiry*. New York: Guilford Press.

Rallis, S. F., Churchill, A., Lawrence, R. B., & Darling, J. (2011). *Supporting professional learning: Impacts of the PLBSS in Portland Public Schools*. Portland, ME; Portland Education Association. Amherst, MA: University of Massachusetts Amherst. https://www.academia.edu/30401313/Supporting_Professional_Learning_Impacts_of_the_PLBSS_in_Portland_Public_Schools

Reyes, A., & Garcia, A. (2014). Turnaround policy and practice: A case study of turning around a failing school with English-language-learners. *The Urban Review, 46*(3), 349–371.

Rogers, P. J. (2008). Using programme theory to evaluate complicated and complex aspects of interventions. *Evaluation, 14*(1), 29–48.Rossman, G. B., & Rallis, S. F. (2010).

Everyday ethics: Reflections on practice. *International Journal of Qualitative Studies in Education, 23*(4), 379–391.
Rossman, G. B., & Rallis, S. F. (2017). *Introduction to qualitative research: Learning in the field*, 4th ed. Thousand Oaks, CA: Sage.
Schön, D. (1983). *Reflective practitioner: How professionals think in action*. New York: Basic Books.
Scott, W. R. (1991). Unpacking institutional arguments. In W. Powell & P. DiMaggio (Eds). *The new institutionalism in organizational analysis*. Chicago, IL: University of Chicago Press, 164–182.
Scott, W. R. (2005). Institutional theory: Contributing to a theoretical research program. *Great Minds in Management: The Process of Theory Development, 37*, 460–484.
Scott, W. R. (2008). Approaching adulthood: The maturing of institutional theory. *Theory and Society, 37*(5), 427–442.
Shaked, H., & Schechter, C. (2017). *Systems thinking for school leaders: Holistic leadership for excellence in education*. Cham, Switzerland: Springer.
Terry, D. (2010). *What's possible: Turning around America's lowest achieving schools*. Ed.gov Blog. Retrieved from https://blog.ed.gov/2010/03/whats-possible-turning-around-americas-lowest-achieving-schools/
Williams, B., & Hummelbrunner, R. (2011). *Systems concepts in action: A practitioner's toolkit*. Stanford, CA: Stanford University Press.

# 3

# SCHOOLS AS SOFT SYSTEMS

## Addressing the Complexity of Ill-Defined Problems

*Sharon D. Kruse*

Approaching school as soft systems suggests that the problems that school leaders face are ill-defined, not easily quantified, and have multiple complex solutions (Jackson, 2003). Relative to technical problems where issues can be defined, objectives and goals set and agreed upon, and results can be measured, soft problems reside in a conceptual space where learning is required to delineate properties, attributes, and phases of action within a larger sphere of concern. Soft problems are complex, wicked problems (Rittel & Webber, 1973), and are not easily defined or resolved. No doubt, technical problems can be complicated to address. They can, and often do, take grit and intellect to solve. When issues and circumstances are complicated, success is relative to the ability of organizational members to define a problem well and to implement practices with fidelity. Moreover, success may prove fleeting, that is, the problem may be solved in the short term, only to have a larger, more pressing, associated problem arise later.

Within the equity realm, a technical problem might be one that asks how students can be provided with school supplies, breakfast and lunch, or transportation beyond that of the usual school day. Such concerns may be addressed by the development of a program, the submission of a grant application, or shifting work assignments for bus drivers or other organizational members. Yet, doing so rarely gets to the roots of equity concerns – the unequal distribution of resources and organizational capital with the school organization. They are soft problems (Jackson, 2003). Those problems are more enduring. As such, a short-term goal might be addressed by any number of programs, yet, larger and more persistent issues remain.

Literature suggests that inequalities in schools fall along racial and socioeconomic lines, including lack of access to experienced teachers, advanced curriculum, lack of funding, discipline disparities, and deficit perspectives about low-income families and students of color (Adamson & Darling-Hammond, 2012; Darling-Hammond, 2004; Gregory, Skiba, & Noguera, 2010; Valencia, 1997). As such, addressing equity within schools and school districts presents problems that do not exist independent of school leaders, teachers or students and their families, are deeply contextual and situational, and open to interpretation. In short, they are complex. Furthermore, when issues of equity have been broadly approached as if they were technical problems, (e.g., establishment of standardized testing, zero tolerance discipline policies) inequities already present with the system (e.g., funding and discipline disparities) have been further exacerbated (Darling-Hammond, 2004; Kruse, Rodela, & Huggins, 2018).

By acknowledging the complexity issues such as equity, school leaders may begin a process of inquiry and learning as part of their equity leadership work. In other words, as leaders strive to understand organizational equity concerns, the focus comes on the investigation of the different value systems, purposeful actions, and facilitation of expressions of the problem/issue situation. In this way, a soft systems approach to organizational work surfaces underlying tensions as they relate to race, class, gender, ethnicity, and difference, and make those tensions a central feature of leadership work.

## Systems Defined

Prior to exploring the ways in which systems theory permeates our thinking about equity school leadership it is worthwhile to theoretically clarify what the word "system" means and how it will be used throughout this chapter. Most of us would suggest that a system is composed of smaller parts that work together to form something new and larger than the sum of its individual parts. As Checkland (1972, 1985, 1999) suggests, in its simplest form, a system is a process or structure that is comprised of smaller, somehow connected things – parts, phenomena, or smaller systems – that come together to form a whole. Yet, absent understanding what "smaller connected things" and "coming together to form a whole" means and entails, the definitions suffer from a lack of clarity and depth. Furthermore, in their simplicity, these definitions obscure the ways in which systems, and in particular school systems, can be quite complex.

Clearly, simple systems exist. When systems have few components and their behavior is understandable and predictable, systems can be thought of as simple. For example, for centuries, a system of taxonomy for the identification of life-forms has endured. The system classifies plants, animals, bacteria, protozoa, and fungi according to kingdom, phylum, class, order, family, genus, and species. By grouping living things by their characteristics (e.g., appearance, reproduction, mobility, and functionality), a determination can be made regarding the identification of a given specimen. While it may be complicated to work with the classification system, and care is needed to ensure each scientific determination is indeed correct, the system itself functions clearly and reliably. Present-day DNA analysis aside, the taxonomy performs much like it did when it was first developed by Carl Linnaeus in 1735. It serves to structure our knowledge about living things and classify them in ways that make our understandings systemized.

Structure or organization are key features of any system. *Organization*, whether through the arrangement of component parts or through the use of clearly defined processes and procedures, is a component of even the simplest of systems. However, no matter how well ordered, as the number of parts within a system grows and the *interconnections* and *interactions* between those parts becomes more involved, the systems themselves grow more complicated. Furthermore, unlike the system of taxonomy where determinations are ultimately binary (i.e., yes/no), systems that require interconnection and interaction between parts are inescapably more complex.

In systems where interconnections between items or events are clear, the outcomes related to those connections are generally causal and linear. In closed systems, systems that are not influenced by their surrounding environment, causal linear interconnections are quite common. For example, the classic assembly line is a *closed system* (Checkland, 1972; Forrester, 1994; Jackson, 2003). Parts move down the line and are assembled in a regular and predictable fashion. They start at one end as separate pieces and emerge as a new whole. Although each subsequent change to the emergent product is the result of an interaction between the worker and the product and each interaction is interconnected as the product moves down the assembly line, daily work occurs insulated from outside factors such as management meetings, sales of product, or even information from other lines within the plant. Additionally, the system of assembly line production is rarely influenced by external forces or factors. Even when the line breaks down or is disrupted, once restarted, work resumes in the same ways it occurred prior to the disruption. In fact, because of their closed system functioning, assembly lines have persisted in manufacturing and other industries for centuries. To this day, they continue to offer the most effective and efficient way to move components down the line and create new product from those parts.

However, as complexity within a system is increased, for example, when external factors influence the workings of a system, the system becomes less predictable and is no longer considered closed. *Complex systems* are considered *open systems* (Ahern, Leavy, & Byrne, 2014a; Checkland, 1981; Lane, 2016) because they are susceptible to external environmental factors that may disrupt organizational work flow, direction, and purpose. Within an open system, feedback from the environment in which the system operates regularly influences the system's productivity, performance, and efficiency. In turn, organizational outcomes cannot be guaranteed nor can they necessarily be causally distinct or resultant from particular, knowable, activities. As Simon (1962) suggests, complex systems are ones, "with many parts interacting richly and interdependently so as to make knowledge and predictions of system-level behavior difficult to derive from knowledge about the behavior of the parts."

Human systems are complex because they are at once never static and contain multiple choice points and decision-making opportunities, all of which are subject to internal and external influences (Ahern, Leavy, & Byrne, 2014b). As such, they are particularly vulnerable. More so, because in open systems vulnerability occurs at multiple levels of the system. These layers include: clarity and agreement concerning organizational *purpose*; how and which *functions* organizational members perform; formation of *relationships* and the processes and power structures by which those relationships are maintained; and the operational *structures* designed to support the work within the organization (Ahern, Leavy, & Byrne, 2014a, 2104b; Gustavsson & Hallin, 2014; Hawkins & James, 2017).

In an ideal or model system, each would flow smoothly into the next with organizational purpose as a central feature informing the functions members would complete, the relationships among members, and the structures needed to support the attainment of valued outcomes and goals. Yet, as we all have experienced, our lives are fraught with problems and the systems in which we work rarely function in ideal ways. Moreover, we all live within the midst of changing and challenging events and incidents that unfold over time and space. Certainly, there are times when we find those events pleasing and we are happy with the results. Yet, more often, events occur in which we find ourselves suggesting that, "something needs to change" or "this could be better only if X or Y happened instead." When we ask those questions of our organizations, we are suggesting that some form of system change occur. We are asking that organizational system interconnections and interactions be altered in some way to produce different, hopefully better, outcomes for members of the organization. Yet, system change is hard.

## Hard Systems Approaches

As Hawkins and James (2017) suggest, the notion of complexity is not new to theorizing in organizational studies. Yet, dating back decades, authors have stressed approaches to organizational change that privilege the logical and rational (Ackoff, 1978; Weick, 1995), and focus on the development of logical structures (Forrester, 1994; Pugh, Hickson, & Hinings, 1969), necessary conditions (Goldratt, 1990; Simon, 1962), and root cause logic (Crawford & Pollack, 2004; Gustavsson & Hallin, 2014) in organizational systems and organizational decision-making theories. Known as *hard systems analysis* (Crawford & Pollack, 2004) these approaches lend themselves to organizational problems and problem situations that are clearly defined, not subject to external influences, and are easily solved. As Heifetz (1994) suggests, this kind of thinking is best utilized when technical problems are present. Reinforcing Heifetz's claim, Simon (1962) and Weick (1995) have long asserted that, in technical work, both the problems to be solved and their solutions are readily located and understood. In this way, technical problems are somewhat mechanical in nature, in that they can be diagnosed, treated, and resolved – in other words, they have the conditions in which hard systems approaches can be effective (Crawford & Pollack, 2004; Rose, 2002).

Hard systems approaches function under the assumption that system goals are clearly defined and do not require further examination or refinement. In this way, hard systems approaches answer "how" questions. For example, "how can this need be met?" or "how should we go about this adoption?" Said another way, hard systems approaches look toward "how to do it" when "what to do" is already well defined. This approach is less good at answering questions that begin with "why" or "what."

This is not to suggest that hard systems approaches are necessarily easy or without challenge. Hard systems work may well be quite complicated, requiring multiple analyses, trials, and evaluations prior to determining a best course of action. Clearly, the elements within a complicated system can be interconnected and multifaceted. Yet, in the end, a complicated problem can be broken down into its component parts and the interconnections between those parts understood and altered as needed (Gustavsson & Hallin, 2014; Hawkins & James, 2017). In short, hard systems approaches serve leaders well when the following conditions are met:

- Goals, objectives, and measures of success are clearly defined and agreed upon;
- Outcomes are expressed in quantitative ways so that success can be unambiguously measured;
- External influences and disruptions to the system are absent or minimal;
- System interventions can be easily monitored and controlled; and
- Effectiveness and efficiency are valued.

Therefore, it is clear that hard systems approaches, while useful, are not well suited to problem situations, like those of equity, that present in complex systems. As has been suggested above, what distinguishes the complex system from that of one that is merely complicated, is the presence of *interaction* between the elements within the system. Moreover, in complex systems the very nature of system interactions changes the system resulting in further complexity and ambiguity (Gustavsson & Hallin, 2014; Hawkins & James, 2017). Something more than hard, rational analysis is needed when these situations present themselves.

## Soft Systems Approaches

In contrast to the research that explores hard systems analysis is the work regarding soft systems approaches or methodology. Soft systems approaches suggest that the problems that school leaders face are ill-defined, not easily quantified, and have multiple complex solutions (Jackson, 2003). The literature (Checkland & Poulter, 2006), has stressed that soft systems approaches are more accurately described as a methodology in that, "it is an ongoing set of principles which can be adapted for use in a way which suits the specific nature of each situation in which it is used" (p. 6). Originating over forty years ago (Checkland, 1972, 1981, 1985, 1999; Checkland & Scholes, 1999) *soft systems methodology* adopts an interpretivist systems perspective suggesting that an organizational member's interpretations of their positional experiences – both positive and challenging – are an integral part of understanding the system itself, and, in turn, influences how problems can be approached. Soft systems methodology favors a more holistic approach to understanding how systems function and includes attention to the ill-defined and ambiguous dynamics within a given system such as a school or school district.

Noting that human decision making rarely follows rational rules, Ahern, Leavy, & Byrne (2014b) suggest that soft systems methodology addresses the limitations of hard systems approaches by acknowledging the multiple and interacting perceptions of reality that exist within human activity systems. In this way, soft systems methodology focuses on surfacing diverse perceptional *worldviews* as part of a process of inquiry within problem-solving and decision-making efforts.

From the equity lens, a longstanding systemic issue in school decision making has been the failure of school leaders to include the voices of youth, parents, and community members, especially those marginalized along lines of race, class, and gender. As Bertrand and Rodela (2017) assert, our conceptions of who are considered leaders in a school setting is overly limited. They note that even within the social justice literature the field often overlooks the leadership of youth, parent, and community members and their contributions to movements towards equity. Yet, historically marginalized individuals and communities have much at stake concerning what goes on in schools and districts and have not been included in the decisions to address inequities they face (Bertrand & Rodela, 2017; Gregory, Skiba, & Noguera, 2010). A soft systems methodology approach pulls those voices into the dialogue changing the locus of control and power from one that rests with individuals with formal school titles and positions (i.e. administrators, teachers) to one that is purposefully more inclusive and shifts existing power dynamics. The purposeful inclusion of alternative voices in the soft systems methodology process changes the decision-making dynamic from one in which equity is a decision-making end (e.g., adoption of programs to serve students and their families) despite overlooking equity in decision making (i.e., who participates in decisions) as the means to this end.

Soft systems methodology is characterized by *purposeful activity* including problem framing from multiple lenses and structured debate about desirable and feasible change. In turn, a key outcome and feature of soft systems methodology is individual and organizational learning. As Železnik, Kokol, and

Vošner (2017) stress, learning and action are fundamental and strategic features of soft systems methodology, defining soft systems methodology as "a *cyclic learning system* which uses models of human activity to explore with the actors in the real-world problem situation, their perceptions of that situation and their readiness to decide upon purposeful action which accommodates different actors' perceptions, judgements, and values" (emphasis added, p. 107).

Other researchers (Forrester, 1994; Rose, 2002; Wang, Liu, & Mingers, 2015) stress that soft systems methodology proceeds from a stance of inquiry when approaching problematic situations. As organizational members jointly research a situation and build rich pictures and interpretive representations of the situation at hand, a variety of viewpoints and approaches to problematic situations are surfaced. In this way, soft systems methodology requires that organizational members jointly enter into problematic situations seeking to understand the complexity of the problem(s) they face. By challenging those within the system not only to think about what they know but, also, to acknowledge the limits of that knowledge – as no representation of a system can be complete –soft systems methodology honors the complexity that is ever-present in systems of human activity.

Surfacing the limitations of our knowledge and belief systems matters because, as Lane (2016) suggests, when faced with complex issues people often do not recognize the limits of their knowledge. This occurs for two significant reasons. First, addressing complex situations and problems is hard work. As a relief from the difficulties that addressing complexity introduces, once we find a convincing cause or explanation for something we stop searching for alternative or additional explanations. We believe we have found an answer and need look no further. Second, our search is often limited by search strategies that favor our deeply held beliefs and biases. We seek confirmation of the very biases we hold. In this way, we are both able to believe we have arrived at a satisfactory conclusion and we have done so in a way that fits our belief structures. In short, our mental models, those abstract sets of ideas that we use to make sense of the world, are often under-informed and positional.

Moreover, information we do not want to acknowledge is often hard to accept. It is simply easier to label for example, a misbehaving student as "a troublemaker" than it is to interpret that behavior as related to a student's need to be noticed, avoid failure, or as a problem with the curriculum being pitched too high or too low to be engaging. In other words, it is easier to believe that problems within the classroom environment are about others (e.g., children and their families), and not us (e.g., teachers, principals, and a system that marginalizes difference). Soft systems methodology seeks to cut through complexity by prompting users to develop new mental models with sufficient richness to inform our problem-solving and decision-making efforts.

Accordingly, soft systems methodology is a purposefully designed four-stage process that addresses problems that require learning rather than more technical or day-to-day issues (see Checkland & Poulter (2006) and Checkland & Scholes (1999) for a detailed account of these methods in use). First, organizational members identify the/a problematic situation in which intervention and change is desired and then develop rich images of the situation from as many organizational members' perspectives as possible. The focus on including a variety of organizational members is designed to establish mutual trust and cooperation in both problem identification and solution finding (Checkland & Poulter, 2006; Checkland & Scholes, 1999; Wang, Liu, & Mingers, 2015). In our consideration of equity, we must accept that developing trust among families and community members, as co-collaborators and partners in the educational system, transcends the usual behavior of schools and school systems. More often, members of the school community who are not part of day-to-day operations (e.g., parents) are excluded from decision-making processes. Therefore, as school leaders take on soft systems methodology as an approach to equity leadership, a focus on developing trust where it has been absent must be an integral part of the work.

As such, focus groups and internal and external community meetings may be more appropriate venues than, for example, individual interviews to collect this data. By holding public events in the community, school leaders have the opportunity to send the message that this work is not business as usual. However, it is important that in these venues, the objective be to actively listen to the viewpoints and experiences of stakeholders. Dependent on the audience, it may be useful to audio or video tape these sessions as well as take extensive and detailed notes assuring attendees that their voices are being accurately and carefully collected. Once assembled, the rich images that arise from multiple descriptions of the issues can then be examined in relation to the following questions. (Note, these items do not comprise an exhaustive list. Instead they are offered as a starting point to foster a discussion about issues of equity, written so as to suggest an approach that favors a stance of inquiry over one of judgement or evaluation.)

- Who are the key players in the situation? Who are the important secondary actors?
- What perspectives do the key players and secondary actors bring to their understanding of the issue(s) at hand?
- How do they describe and define the situation and its core issues?
- What is the cultural environment of the situation? In what ways do different groups (e.g., teachers, parents, community members, students) frame the issues?
- In particular, what norms and values shape the situation and in what ways do they differ among stakeholder groups?
- How is the situation affected by politics or power relationships?
- What sources of power are important within and to the situation?

- How is the situation described by each stakeholder group? In what ways are the descriptions similar? In what significant ways do they differ?
- What has been learned by unpacking the issues and what might still be missing from this problem-solving knowledge base?

Following the development of a rich picture of the variety of ways the problem situation can be described, participant decision makers are encouraged to select several representative perspectives for further deliberation and build a conceptual model (e.g., a graphic or written representation) of the problem situation within system. Such models need not be extensive or particularly artistic. In fact, they are generally not for public viewing or use. Instead they serve to allow decision makers the opportunity to view the issues in a tangible and intentional fashion, creating a clear and agreed upon conception of the issue/problem for further analysis. However, dependent on the issue at hand it may be useful to vet these models with internal and external stakeholder groups to be sure their worldviews are represented accurately. In this way, soft systems methodology stretches our definition of distributed or shared leadership, repositioning individuals so that previously marginalized members can work together as leaders to enact change.

Third, participant decision makers are encouraged to develop a conceptual model of the change system (e.g., potential actions to initiate change efforts) and compare that model to the real-world situation as described by the stakeholders. In this iteration of the work, organizational members are looking to answer questions such as:

- What reasonable limitations are present that bound our choices for activity and change?
- How would each stakeholder group respond to the planned activities and changes?
- What are the benefits to each stakeholder group if a particular approach is adopted?
- Do planned changes respond to the issues and concerns presented by each stakeholder group? If not, where are the potential areas of disagreement and resistance?
- If one stakeholder group is privileged, why was that the case and how will those choices be explained?
- What are the potential pitfalls of this plan? What might some unintended consequences of these actions be?
- How might resistance, disagreement, and challenge be addressed?

Here again, an opportunity to advance equity in schools presents itself as after due consideration of the problem situation at hand. As the consequences of the variety of decision models are considered, decision makers should move to action in ways that are responsive to the broad range of issues posed. In turn, organizational members must mindfully attend to how changes (e.g., new policies or practices) are received and implemented, and ultimately, if the changes remedy the original problem situation in meaningful ways. In other words, the soft systems methodology model provides school leaders with an opportunity to engage in work that demonstrates the practice of inclusion in the process of thinking about how schools can become more equitable.

By building on the interplay of "doing in order to think" and "thinking in order to do" (Mintzberg, 2009, 2013), soft systems methodology focuses attention on dynamic organizing (e.g., conceptual modeling and consideration from broad and inclusive perspectives) and organizational learning (e.g., developing deep understanding) designed to foster equitable problem-solving efforts. Soft systems methodology suggests that only by facilitating organizing and learning together, can ill-defined and ambiguous problems be addressed. Therefore, soft systems methodology is best employed when:

- Goals, objectives, and measures of success are ambiguous and may have several definitions;
- Outcomes may be expressed in qualitative ways and differ dependent on stakeholder group;
- The problem situation is highly subject to external influences;
- System interventions require the exploration of a variety of alternative solutions; and
- Values, relationships, cultural difference and experience, and meaning are valued.

In conclusion, soft systems methodology is an approach that is oriented toward learning in problematical and ill-defined problem situations and addresses questions that ask "why" rather than "how." It favors situations that assume the world can be explored through an inquiry lens and that the development of understanding concerning others' worldviews is a necessary component of both understanding the issue as well as evaluating potential solutions. Yet, soft systems methodology also assumes that inquiry is never-ending and that even as one solution is launched, other issues will surface that will, in turn, influence and inform next steps and new ways forward toward better organizational choices and leadership.

## Soft Systems Methodology as a Form of Holistic Organizational Learning

A hallmark of soft systems methodology, as opposed to, for example, action research (Carr & Kemmis, 1986), is its focus on inquiry into the perspectives of a variety of organizational members within and related to the organizational system.

This is not to say that action research precludes such data collection. Rather, it is to emphasize that soft systems methodology requires the collection of multiple worldview portraits as a method of deepening understanding regarding ill-defined and ambiguous problem situations. Knowledge that is developed within soft systems methodology processes of inquiry necessitates that organizational insiders look beyond that which they know to include the contextual and situational knowledge that others bring to the problem-framing and -solving enterprise. In turn, by acknowledging that knowledge about any given situation is dispersed among the many, a conceptual holistic model of the issue(s) may be drawn.

Such work, inescapably, leads to the development of new understandings across the landscape of the problem-situation. In this way, soft systems methodology is more like jazz improvisation than a technically perfect rendering of a classical piece. In the playing of classical music, the music is known in advance and the musicians play exactly what is written on the page, whereas in improv the major features of a piece are known; however, the musical creation is spontaneous and inspired by listening and responding to the contributions of the other musicians. Jazz improv emphasizes the contextual and is produced by the collective efforts of the whole, who deliver the piece while they create it. Producing music in this way is a complex process. Even as polished musicians bring to the effort their prior knowledge and skill, improv requires that they learn their way into each night's performance by attending to what others play and do.

In this way, the band learns what comes next and creates the experience, *in the context of delivering* the performance. By creating as they go, the process is more like "becoming" (e.g., discovering, gaining new insights) than "being" (e.g., replicating that which is already known) (Lave & Wenger, 1991; Wenger, 2001). As a way of dealing with complex problems, this notion of learning emphasizes the emergent and experiential. It highlights how multiple perspectives increase our ability to bring knowledge and action together in practice (Argyris, 1977; Argyris & Schon, 1996).

However, unlike jazz, where performance outcomes are short-lived and passing, organizational learning requires attention to the ways new ideas become adopted and employed with the system. Building on experience as a fundamental mechanism to foster learning, organizational learning relies on members' ability to adjust current behavior based on past performance and in light of current goals. Additionally, organizational learning highlights how the absence of comprehensive knowledge in any problem situation is an intrinsic characteristic of complexity in organizations (Gustavsson & Hallin, 2014; Weick, 1995). In this way, soft systems methodology creates avenues to more fully develop a holistic picture of how the current organization is experienced by all members and why it may be hindered in achieving its goals (Checkland & Poulter, 2006; Checkland & Scholes, 1999).

Using an organizational lens, soft systems methodology shifts the emphasis for learning, from "what can we do?" to an orientation that first asks "why are we experiencing this problem and how do others view it?" Accordingly, attention is paid to comprehensively learning the boundaries of an issue prior to making any attempt to resolve it. In this form, an organization's learning is more about identifying and confronting older, unproductive, organizational routines and less about importing new policies, practices, or procedures absent a thorough examination for organizational fit. Organizational fit is an important theme in thinking about soft systems methodology as a process of organizational learning because of the stress soft systems methodology places on *mutual interest* (Ahern, Leavy, & Byrne, 2014a, 2104b; Argyris & Schon, 1996).

Remember that complex problems are, in part, complex because they are not easily defined. This lack of specificity confounds our ability to locate robust solutions because absent a fulsome definition of the issues that face us our solutions can only be partial. Obtaining a fulsome definition of the problem situation requires that learning transcend that of an individual – or even a small team. It requires that learning be vigorously pursued by the many and that it be in service of the communal and mutual interest (Wenger, 2001). As such, promoting work that engages others in common foci directed toward challenging mutual goals and managing work toward those goals are two complementary and crucial ingredients for success in complex environments (Ahern, Leavy, & Byrne, 2014a; Heifetz, 1994).

The first, engaging others in common foci directed toward mutual goals, compels leaders to think about what motivates their colleagues, external partners, and internal stakeholders by attending to the values, beliefs, and aspirations that encourage them to become part of communal efforts. Beyond the symbolic and cultural value of inspiring others to contribute to the whole, doing so minimizes conflict caused by incompatible goal alignment, poor communication around core ideas or ideals, or unfair or unclear expectations (Checkland, 1985, 1999; Crawford & Pollack, 2004). The second, managing work toward those goals, necessitates pacing work in explicit ways.

As Mintzberg (2009, 2013) notes, when people understand what is broadly expected (e.g., treating each other with respect), they are better able to maintain commitment to the common good as well as deepen their own sense of contribution toward communal efforts. In this way, community members can learn their way into new ways of organizing and being. As a result, the organization can change how it defines itself and, in turn, create meaningful changes in its organizational culture (Kruse & Johnson, 2017; Kruse & Louis, 2009). Furthermore, by purposefully broadening who is considered a full member and respecting their contribution(s) in the problem resolving domain, this approach can support lasting and rich organizational change and increased equity.

## Conclusions

There is, of course, a consequence of drawing a hard and fast distinction between hard and soft systems theorizing as it relates to organizational learning and systems thinking more generally. It is worth problematizing here. As with any dichotomy, unnatural distinctions and boundaries are drawn for the sake of definitional clarity. Yet, in doing so the in-between is ignored. Clearly, systems, like those found in schools, require both hard systems work and soft systems work. Inasmuch as equity serves as an exemplar for the ways in which soft systems methodology can provide a way to unpack the positionalities of internal and external school community members, soft systems methodology is not well-suited to projects that lack complexity. In fact, turning all problem situation work in a soft system project risks a form of organizational paralysis where all too much data and input is required for work to be efficiently or effectively completed. Similarly, there are certainly projects that benefit from inputs that draw from hard and soft approaches.

Yet, soft systems methodology cannot be dismissed as overly burdensome or time consuming or as untenable within any given school community. In fact, dismissing the methodology out of hand risks avoiding the very issues that reinforce systemic inequality and inequity. Among them are issues of power and status and who benefits from upholding existing power structures and organizational constructions (Bertrand & Rodela, 2017; Kruse, Rodela, & Huggins, 2018; Valencia, 1997). When we rely solely on individually held and therefore subjective explanations for the defense of existing organizational practices, we increase the social power of those who hold those positions and explanations. Moreover, subjective explanations (e.g., "it's policy" or "it's how we've always done things") include not only one's individual knowledge concerning "how schools should work" but also their biases and prejudices (e.g., "it's always those kids"), their self-deception (e.g., "we don't have issues with equity"), and their self-censorship (e.g., "I can't possibly admit this). In both cases – self-deception and self-censorship – one's perceptions of the power relationships between themselves and others within the school community, as well as their perceptions about the consequences of conforming to or breaking from expected behaviors or opinion, color their understanding of problem situations and subsequent decision actions (Cooper, 2003). Thus, one of the key tasks for addressing equity is to find ways to uncover systemic distortions within our understandings of "how school is done" and our self-deceptions of our role within those systems. Soft systems methodology provides us with a tool to dive widely and deeply into understanding not only our own stance toward equity, but that of important and valued others.

## References

Ackoff, R. L. (1978). *The art of problem solving*. New York: Wiley.

Adamson, F., & Darling-Hammond, L. (2012). Funding disparities and the inequitable distribution of teachers: Evaluating sources and solutions. *Education Policy Analysis Archives, 20*. http://epaa.asu.edu/ojs/article/view/1053

Ahern, T., Leavy, B., & Byrne, P. J. (2014a). Complex project management as complex problem solving: A distributed knowledge management perspective. *International Journal of Project Management, 32*, 1371–1381.

Ahern, T., Leavy, B., & Byrne, P. J. (2014b). Knowledge formation and learning in the management of projects: A problem solving perspective. *International Journal of Project Management, 32*, 1423–1431.

Argyris, C. (1977). Double loop learning in organizations. *Harvard Business Review, 55*(5), 115–125.

Argyris, C., & Schon, D. A. (1996). *Organizational learning: Theory, method, and practice.* Reading, MA: Addison-Wesley.

Bertrand, M., & Rodela, K. C. (2017). A framework for re-thinking educational leadership in the margins: Implications for social justice leadership preparation. *Journal of Research on Leadership Education*, https://doi.org/10.1177/1942775117739414

Carr, W., & Kemmis, S. (1986). *Becoming critical: Education, knowledge and action research.* Abingdon, UK: Deakin University.

Checkland, P. (1972). Towards a systems-based methodology for real-world problem solving. *Journal of Systems Engineering, 3*(2), 87–116.

Checkland, P. (1981). *Systems thinking, systems practice.* Chichester, UK: John Wiley and Sons Limited.

Checkland, P. (1985). From optimizing to learning: A development of systems thinking for the 1990s. *Journal of the Operational Research Society, 36*(9), 757–767.

Checkland, P. (1999). *Systems thinking, systems practice* (2nd ed.). Hoboken, NJ: John Wiley and Sons Limited.

Checkland, P., & Poulter, J. (2006). *Learning for action: A short definitive account of soft systems methodology and its use of practitioners, teachers, and students.* Hoboken, NJ: John Wiley and Sons Limited.

Checkland, P., & Scholes, J. (1999). *Soft systems methodology in action.* Hoboken, NJ: John Wiley and Sons Limited.

Cooper, T. (2003). "Critical Management, Critical Systems Theory and System Dynamics" at The Third International Conference of Critical Management Studies, published in *Critique and inclusivity: Opening the agenda.* Lancaster, UK. http://www.mngt.waikato.ac.nz/ejrot/cmsconference/2003/proceedings/orsystems/cooper.pdf

Crawford, L., & Pollack, J. (2004). Hard and soft projects: A framework for analysis. *International Journal of Project Management, 22*, 645–653.

Darling-Hammond, L. (2004). The color line in American education. *Du Bois Review, 1*(2), 213–246.

Forrester, J. W. (1994). System dynamics, systems thinking, and soft OR. *System Dynamics Review, 10*(2–3), 245–256.

Goldratt, E. M. (1990). *The haystack syndrome.* Croton-on-Hudson: North River Press.

Gregory, A., Skiba, R. J., & Noguera, P. A. (2010). The achievement gap and the discipline gap: Two sides of the same coin. *Educational Researcher, 39*(1), 59–68.

Gustavsson, T. K., & Hallin, A. (2014). Rethinking dichotomization: A critical perspective on the use of "hard" and "soft" in project management. *International Journal of Project Management, 32*, 568–577.

Hawkins, M., & James, C. (2017). Developing a perspective on schools as complex, evolving, loosely linking systems. *Educational Management Administration and Leadership.* DOI: 10.1177/1741143217711192

Heifetz, R. (1994). *Leadership without easy answers.* Cambridge, MA: Harvard.

Jackson, M. C. (2003). The origins and nature of critical systems thinking. In G. Midgley (Ed.), *Systems thinking* (Vol. 4, pp. 227–253). London: Sage.

Kruse, S. D., & Johnson, B. L. (2017). Tempering the normative demands of PLCs with the organizational realities of life in schools: Exploring the cognitive dilemmas faced by educational leaders. *Educational Management, Administration, & Leadership, 45*(4), 588–604.

Kruse S. D., & Louis, K. S. (2009). *Building strong school cultures: A leader's guide to change.* Thousand Oaks, CA: Sage.

Kruse, S. D., Rodela, K. C., & Huggins, K. S. (2018). Messy messages and making sense across complex contexts: A regional network of superintendents confronting equity. *Journal of School Leadership, 28*(1).

Lane, D. C. (2016). "Till the muddle in my mind have cleared Awa": Can we help shape policy using systems modeling? *Systems Research and Behavioral Science, 33*(5), 633–650.

Lave, J., & Wenger, E. (1991). *Situated learning: Legitimate peripheral participation.* Cambridge, UK: Cambridge University Press.

Mintzberg, H. (2009). *Managing.* San Francisco, CA: Berrett-Koehler Publishers.

Mintzberg, H. (2013). *Simply managing: What managers do and can do better.* San Francisco, CA: Berrett-Koehler Publishers.

Pugh, D. S., Hickson, D. J., & Hinings, C. R. (1969). An empirical taxonomy of structures of work organizations. *Administrative Science Quarterly. 14*(1), 115–126.

Rittel, H., & Webber, M. (1973). Dilemmas in a general theory of planning. *Policy Sciences, 4,* 155–169.

Rose, J. (2002). Interaction, transformation, and information systems development: An extended application of soft systems methodology. *Information Technology and People, 15*(3), 242–268.

Simon, H. (1962). The architecture of complexity. *Proceedings of the American Philosophy Society, 100*(62), 467–482.

Valencia, R. (1997). *The evolution of deficit thinking: Educational thought and practice.* New York: Routledge Farmer.

Wang, W., Liu, W., & Mingers, J. (2015). A systemic method for organizational stakeholder identification and analysis using soft systems methodology. *European Journal of Operational Research, 246,* 562–574.

Weick, K. (1995). *Sensemaking in organizations.* Thousand Oaks, CA: Sage.

Wenger, E. (2001). *Communities of practice: Learning, meaning, and identity.* Cambridge, UK: Cambridge University Press.

Železnik, D., Kokol, P., & Vošner, H. B. (2017). Adapting nurse competence to future patient needs using Checkland's soft systems methodology. *Nurse Education Today, 48,* 106–110.

# 4

# PRINCIPALS' SYSTEMS THINKING

## The Meaning and Measure of a Leadership Construct

*Haim Shaked, Pascale Benoliel, Nechama Nadav, and Chen Schechter*

A school is an inherently complex organization, involving a vast number of interacting functions, people, and purposes (Crick, Barr, Green, & Pedder, 2017; Ng, 2015). As such, school leaders often encounter conflicts concerning potential choices of action, each with its pros and cons, as well as conflicts emanating from various stakeholders' diverse or even incompatible aspirations, views, expectations, and demands (Saiti, 2015). Practitioners and researchers alike agree that recent years have brought even more challenging complexities to school leadership (Fullan, 2014; Hargreaves & Braun, 2013). Hence, the traditional "toolkit" that school principals have at their disposal is often insufficient to provide effective strategies for meeting today's almost incredible array of difficulties and expectations (Louis, Leithwood, Wahlstrom, & Anderson, 2010). Therefore, school principals sorely need comprehensive theoretical frameworks accompanied by complementary strategies to guide them toward success (Fullan, 2014). Particularly, current school principals facing today's educational leadership complexities may benefit from the holistic perspective of systems thinking.

Systems thinking (ST), which does not try to break systems down into parts in order to understand them but rather concentrates on how the parts function together in networks of interaction (Gharajedaghi, 2011; Senge, 2006), has been recognized as an effective management approach (Brown, 2012; Jolly, 2015; Wilson & Van Haperen, 2015) with a wide range of applications (e.g., Holmes, Finegood, Riley, & Best, 2012; Leischow et al., 2008; Tejeda & Ferreira, 2014). For example, strong statistical correlations have been found between ST and project performance (Elm & Goldenson, 2012).

However, in the context of school leadership, ST has not yet received sufficient empirical attention (Shaked & Schechter, 2017). Several educational guidebooks have suggested ways to implement ST into schools, offering practical advice on the use of such thinking to confront present-day educational demands and challenges, including structured models for successful educational reforms (e.g., Fullan, 2005; Senge et al., 2012). Yet, few researchers have empirically examined the dimensions of principals' ST (PST) and their application in school practices. Based on our recent conceptualization of ST in school leadership and qualitative exploration of PST characteristics (Shaked & Schechter, 2014, 2017), in this chapter we report on our development of a valid, reliable instrument to measure PST as it is manifested in today's schools. Before describing the scale's development and testing, we first present theoretical background on ST in general, and ST in the context of school principals, and PST's expected associations with two other relevant, established constructs: instructional leadership and organizational commitment.

## The Systems Thinking (ST) Framework and Definitions

ST is not a discipline in and of itself, but rather an interdisciplinary conceptual framework that is employed in a wide range of areas. It may be regarded as a type of orientation or approach toward the world at large, or a model for thinking and learning about any given system – scientific, organizational, personal, or public (Cabrera & Cabrera, 2015). Thus, the literature on ST encompasses a broad range of fields, yielding a variety of definitions for the concept. Mainly representing the interdisciplinary area of systems science, these definitions cover complex systems, cybernetics and dynamical systems theory, and applications in the natural and social sciences as well as in engineering (Hieronymi, 2013).

Here are some of the definitions and explanations for ST formulated by scholars in recent decades, presented in chronological order: Senge (1990, p. 68) defined ST as "a discipline for seeing wholes. It is a framework for seeing interrelationships rather than things, for seeing patterns of change rather than static 'snapshots.' It is a set of general principles. . . It is also a set of specific tools and techniques." Richmond (1994, p. 141) asserted that ST is "the art and science of making reliable inferences about behavior by developing an increasingly deep understanding of underlying structure." Sterman (2000, p. 4) viewed ST as "the ability to see the world as a complex system, in which we understand that 'you can't just do one thing' and that 'everything is connected to everything else.'" Arnold and Wade (2015, p. 675) opined that ST is "a set of synergistic analytic skills used to improve the capability of identifying and understanding systems, predicting their behaviors, and devising modifications to them in order to produce desired effects. These skills work together as a system."

Despite the absence of one commonly accepted definition for ST, these diverse definitions clearly yield two basic complementary meanings of the construct: rising above the separate components to see the whole system, and thinking about each separate component as part of the whole system (Shaked & Schechter, 2014, 2017). Thus, ST advocates stepping back to focus not only on the trees but also on the forest – where the trees are particular situations or limited domains, while the forest is an overall view (Richmond, 2000). From the ST perspective, people should not get "snagged" by the details. Instead, one must discern an overall pattern from the mass of detail, aggregating and integrating this data into a holistic framework (Frank, 2012). The only way to fully understand a system is to understand its parts in relation to the whole, because the system's defining characteristics comprise the characteristics of that very whole, which cannot be found in the isolated parts. Once the system is analyzed – taken apart – these defining characteristics of the whole get lost (Jackson, 2010).

According to ST, the many interactions that transpire within each system cannot be reduced to a single cause-effect relationship. The first event is likely to contribute to the second event, but many additional events are also seen as contributing to that second event. Thus, all explanations for the second event should take into account the influence of multiple factors. The explanation for any phenomenon within a system cannot be too simplistic, and the impact of the environment cannot be ignored (Hitchins, 2007). That is, causation in systems is less straightforward and salient, necessitating a broader understanding of the system as a whole (Senge, 2006).

ST is considered to be a key management competency (Brown, 2012). Especially in the twenty-first century, as the world becomes ever more tightly interwoven globally and the pace of change continues to grow, managers need to become increasingly "system-wise," applying ST principles and practices in their organizations (Jolly, 2015; Wilson & Van Haperen, 2015). The benefits of ST in various management areas (e.g., Holmes et al., 2012; Leischow et al., 2008; Tejeda & Ferreira, 2014) suggest that ST may also be valuable for school principals.

## *Characteristics of School Principals' Systems Thinking (PST)*

The existing literature on ST in school leadership, which we recently reviewed (Shaked & Schechter, 2017), is important, albeit limited. To advance knowledge in this domain, we conducted a qualitative study (Shaked & Schechter, 2014, 2017) to pinpoint the characteristics of PST. This study identified four major ways in which principals apply the ST approach and perform at the ST level in their schools:

(1) **Leading wholes.** This characteristic of "leading wholes" refers to principals' holistic perspective, oriented toward seeing the big picture and not only its individual parts. Such principals perceive and conceptualize all aspects of school life as one large system. Sub-characteristics include, for example, the capacity to tolerate ambiguity and uncertainty, which derives from an ability to understand that a current situation's seemingly separate or confusing details are actually parts of one big picture.

(2) **Influencing indirectly.** This characteristic refers to principals' ability to address school goals and challenges circuitously, based on their awareness that countless reciprocal influences are at play among various school elements, each of which is connected to others, affecting them and being affected by them. Sub-characteristics include, for example, the ability to motivate teachers through responsibility rather than through blame, which derives from principals' recognition of the indirect influences of their own actions on the staff's functioning.

(3) **Adopting a multidimensional view.** This characteristic refers to the principal's contemplation of several aspects of a given issue simultaneously. Any given issue's emergence and existence will be regarded by effective principals as stemming from a wide range of potential sources. They will consider the issue's consequences, before proceeding to predict various optional courses of future action or development. Sub-characteristics include, for example, openness to a variety of opinions, which enables seeing the multiple aspects of any situation within school.

(4) **Evaluating significance.** This characteristic refers to the principal's ability to envision elements of school life according to their significance for the entire system. Principals prioritize issues to be resolved according to their degree of importance and urgency, while also identifying patterns as they go along. Sub-characteristics include, for example, the capacity to balance internal and external relationships and needs, thereby successfully negotiating between internal desires and capacities and outside demands and expectations.

These four characteristics clearly tie in with the two major meanings of ST. Leading wholes and adopting a multidimensional view reflect *seeing the whole beyond its parts*, while influencing indirectly and evaluating significance are related to *seeing the parts in the context of the whole*.

To date, ST in school leadership was rarely explored empirically, and then only through qualitative methodology, which did not yet yield a valid instrument to measure this construct and distinguish its characteristics. The present chapter presents our study that aimed to design such an instrument in line with the theoretical model and structural characteristics of PST, while testing its criterion-related validity by examining the newly developed scale's correlations with two established, related constructs: instructional leadership and organizational commitment.

## *Systems Thinking (ST), Instructional Leadership (IL), and Organizational Commitment (OC)*

In recent decades, principals have faced superintendents' and policy makers' clear expectation to become "instructional leaders" – who focus efforts squarely on the improvement of teaching and learning as the school's primary goal (May & Supovitz, 2011; Rigby, 2014). The instructional leadership (IL) framework arose from the close connections identified between teachers' quality of instruction and students' academic results (Murphy, Neumerski, Goldring, Grissom, & Porter, 2016). Thus, as instructional leaders, principals are expected to give student learning top priority, with everything else revolving around the enhancement of that learning (Hallinger, 2011; Supovitz, Sirinides, & May, 2010). Practically, IL reflects those actions that a principal takes to promote growth in student learning and academic success (Armstrong, 2007; Stronge, Richard, & Catano, 2008; Walker & Slear, 2011). The links between principals' IL and students' achievements, predominantly through indirect means, have been clearly established through research (Glickman, Gordon, & Ross-Gordon, 2014).

IL may be related to PST because, as an approach that allows a view of the whole picture, ST enables managers to concentrate on the organization's main goals (Frank, 2012). It allows managers to keep focused on the organization's key objectives, and not to be distracted by other events or pressures (Brown, 2012; Jolly, 2015). Therefore, PST may be linked to IL, which calls upon principals to maintain their focus on achieving the school's central goal – to improve teaching and learning in order to affect student achievements (Neumerski, 2012).

In addition, the current study explored the relations between the characteristics of PST and teachers' organizational commitment (OC). OC as defined by Mowday, Steers, and Porter (1979) is "the relative strength of an individual's identification with and involvement in a particular organization" (p. 226). ST was found to be advantageous for various managerial aspects (Holmes et al., 2012; Tejeda & Ferreira, 2014), including increased OC among employees. Specifically, in schools, the OC of teachers is important for school effectiveness and indirectly affects students' outcomes (Park, 2005). The OC concept is based on three factors: *identification* – acceptance of the organization's goals and values; *involvement* – willingness to invest effort on behalf of the organization; and *loyalty* – the importance attached to keeping up membership in the organization (Bogler & Somech, 2004). Several studies indicated that teachers' commitment to the school can be an important predictor of teachers' job performance because it is positively related to their dedication to attaining organizational goals (Dee, Henkin, & Singleton, 2006; Sammons et al., 2007). When a manager is running at systems level, employees' psychological attachment to the organization is expected to increase (Brown, 2012; Jolly, 2015). Thus, PST may be related to teachers' OC.

## Development of the PST Scale

To empirically assess the characteristics of PST, we sought to identify the practical ways in which principals might lead schools through the ST framework, as perceived by the mid-level school leaders working in everyday situations under the principal. Mid-level leaders constitute the "intermediate layer" in the school's organizational structure, located between the senior leadership – which usually includes the principal and assistant principal – and the classroom teachers (Crane, 2014; Gurr & Drysdale, 2013). Inasmuch as they hold management responsibility either for staff or for a certain aspect of the school's work, such as grade-level coordinators or subject coordinators, their role involves close work with principals (Fleming & Amesbury, 2012; Thorpe & Bennett-Powell, 2014). Thus, they were chosen as reliable informants to reporting on the everyday ST of their principals.

As described below, we developed the Principal System Thinking (PST) Scale in two phases, examining three independent samples. In Phase 1, we first generated relevant items for the PST Scale based on our prior qualitative research (Shaked & Schechter, 2014, 2017). These potential items were then pooled, subjected to content validation, and pilot-tested on 185 mid-level elementary school leaders using Principal Components Analysis to explore the scale's factor structure, reliability, and validity. In Phase 2, a revised scale was created, the items were submitted to exploratory factor analysis ($N = 129$), and then the proposed scale was validated by confirmatory factor analysis ($N = 100$). Then, we examined correlations of PST with the two other established constructs, IL and OC, to check criterion-related validity.

### *Phase 1: Item Generation, Content Validation, and Pilot Testing (N = 185)*

**Item generation.** To design a measure based on Shaked and Schechter's (2014, 2017) aforementioned four characteristics of ST in school leadership, we have developed empirical indicators specifically formulated for mid-level school leaders to assess practical indicators of their principals' competencies. Thus, we initially designed 45 items with the following characteristic-level distribution: Leading wholes (14 items); Influencing indirectly (7 items); Adopting a multidimensional view (18 items); and Evaluating significance (6 items).

**Content validation.** Proposed items' applicability was evaluated by five experts from the educational administration field and two additional experts from the ST field. Experts were asked to evaluate the 45 proposed PST Scale items, and to either retain or generate a modification/replacement for each item. Thereafter, each member of our research team independently reviewed the full list of potential items and the experts' comments. Based on this procedure, a draft version of the PST Scale was devised.

Next, 60 teachers from various elementary schools in Israel evaluated the draft version of the proposed Hebrew scale. Their remarks regarding items' clarity and response scale, as well as their recommendations for modifying, adding, and deleting items, were considered. This procedure yielded a 38-item pilot scale to be rated along a 5-point Likert-type response scale ranging from *Never* (1) to *Always* (5). The pilot PST Scale had the following subscale distribution according to the four characteristics: Leading wholes (13 items); Influencing indirectly (5 items); Adopting a multidimensional view (15 items); and Evaluating significance (5 items).

**Pilot study: Item reduction (Principal Components Analysis).** The 38-item pilot scale was distributed to a sample of 185 mid-level leaders (response rate = 70%) from 43 randomly selected elementary schools in Israel. These mid-level leaders had a mean age of 38.30 years (*SD* = 8.01), 50.3% were women, and their mean job tenure in the profession was 12.32 years (*SD* = 7.88). Participation was voluntary, the purpose of the study was explained, anonymity was guaranteed, and the importance of candid answers was emphasized.

Data collected from these 185 mid-level leaders were subjected to principal components factor analysis with varimax rotation of the items, identifying common factors with eigenvalues greater than 1. In general, factor loadings were strong, above .50. Seven items, which loaded on more than one factor at .40 or above, and six items which loaded lower than .35 on any one factor, were eliminated. Moreover, four items that substantially reduced subscales' internal consistency (Cronbach coefficient alpha) were also deleted (Items 3, 10, 11, 12). The weightings of Items 14, 16, 19, 29, 35, 36, and 37 could not be classified under any of the categories; thus, these items were omitted from the final scale. Items 5, 15, 24, 26, 33, and 38 were removed in order to resolve some of the multicollinearity between factors and to improve alphas. Finally, one factor was judged problematic (Items 1, 2) and thus was classified as non-interpretable because no single grouping of items representing the intended factor emerged as the dominant source for the factor.

Thus, altogether, a total of 19 items were eliminated during this procedure. Deleted items included: "The principal tends to react to anything that happens in the school, big or small; The principal takes into account that action in one realm indirectly affects others; When the principal tries to understand why a certain event occurred in school, he/she usually come up with only one explanation for it." This process resulted in a set of 19 items to measure PST, entering Phase 2.

## *Phase 2: Exploratory Factor Analyses (N = 129), Confirmatory Factor Analyses and Criterion Validity (N = 100)*

**Exploratory factor analysis.** We conducted this analysis to explore the number of factors that account for covariation between variables due to insufficient a priori evidence to form a hypothesis about the number of factors underlying the data (Stevens, 1996). We distributed the 19-item scale to a new sample of 129 mid-level leaders (response rate = 90%) from 42 new randomly selected elementary schools. This sample was 63 percent female, with a mean age of 37.79 years

($SD$ = 6.23) and mean job tenure in the profession of 11.33 years ($SD$ = 7.03). As before, participation was voluntary, and we explained the study's purpose, guaranteed anonymity, and emphasized the importance of candid answers.

Exploratory factor analysis of the item matrix was performed to study which items clustered together and which did not. For this purpose, a principal-axis factor analysis, rotated using Kaiser's (1958) varimax criterion, was used to examine the 19-item measure and to identify common factors with eigenvalues greater than 1. Items 8 and 17 were removed in order to resolve some of the multicollinearity between factors and to improve alphas. This analysis enabled us to discern a 4-factor structure among the remaining 17 items, which explained 56.91 percent of the variance. As seen in Table 4.1 (which presents rotated factor loadings for each of the 17 items), all four factors had eigenvalues greater than 1 (ranging from 5.10 to 1.33). After examining the items that loaded on each factor, the four characteristics of PST were identified: Evaluating significance (4 items; $\alpha$ = .75); Openness to a variety of opinions (4 items; $\alpha$ = .78); Leading wholes (3 items; $\alpha$ = .66); and Adopting a multidimensional view (6 items; $\alpha$ = .70). The PST Scale demonstrated a high level of internal consistency ($\alpha$ = .85).

**Confirmatory factor analysis.** This procedure used AMOS 21.0 structural equation modeling software to validate the 17-item PST Scale, as administered to a new independent sample of 100 elementary school mid-level leaders. This sample size is considered appropriate for confirmatory factor analysis model testing because the ratio of sample size to items surpassed the minimum of 5:1 (Hair, Anderson, Tatham, & Black, 1998). This third sample was 62% female, with mean age of 40.06 years ($SD$ = 8.30) and mean job tenure in the profession of 13.49 years ($SD$ = 7.97). As before, participation was voluntary, and we explained the study's purpose, guaranteed anonymity, and emphasized the importance of candid answers.

This analysis was based on the covariance matrix, and we used maximum likelihood estimation, yielding the results shown in Table 4.2. Fit indices provide feedback about the appropriateness of hypothesized research models based on the covariance structure of the observed data. As recommended by researchers (Jöreskog & Sörbom, 1993; Kline, 1998), we used several goodness-of-fit indices: the chi-square statistic divided by the degrees of freedom ($\chi^2$/df) is recommended to be less than 3; the values of comparative fit index (CFI), incremental fit index (IFI) and Tucker-Lewis index (TLI) are recommended to be greater than.90; root mean square error of approximation (RMSEA) is recommended to be up to.05, and acceptable up to.08 (Steiger, 1990).

We compared four models: the proposed second-order model (Model 1) and three first-order models (Models 2 to 4). Model 1 denotes that the 17 observable items load on the four first-order factors (Adopting a multidimensional view; Openness to a variety of opinions; Leading wholes; and Evaluating significance), which in turn load on the second-order factor (PST). The specific PST

**TABLE 4.1** Item Loading Values from Exploratory Factor Analysis of the Principals' Systems Thinking Scale (N = 129)

| | *Factors* | | | |
|---|---|---|---|---|
| | *Evaluating significance* | *Openness to a variety of opinions* | *Leading wholes* | *Adopting a multi-dimensional view* |
| The principal spots small events that are nevertheless meaningful | .82 | | | |
| The principal tends to take unexpected occurrences into account | .62 | | | |
| The principal explains decision-making processes in accordance with the school's goals | .76 | | | |
| The principal tends to take different points of view into consideration when deciding on various matters | .55 | | | |
| The principal tends to involve teachers in decision-making processes | | .83 | | |
| The decisions made by the principal are based on the school's policy | | .79 | | |
| The principal usually tries to figure out how various events have led to each other | | .71 | | |
| The principal engages in dialogue with those holding educational outlooks that differ from his/her own | | .57 | | |
| The principal tends to suggest solutions that affect the entire work environment, rather than just specific details | | | .79 | |
| The principal attempts to identify repetitive patterns in the information at hand | | | .67 | |
| At meetings and discussions, the principal tries to present most points of view | | | .66 | |
| The principal functions well also in ambiguous situations | | | | .75 |
| During decision-making, the principal tends to view the entire picture before examining its details | | | | .46 |
| The principal takes into account that the effects of a certain action may vary in different situations | | | | .48 |
| The principal acts with the understanding that small changes can bring about significant results | | | | .59 |
| The principal tries to understand how certain components of the school affect the way other components in it function | | | | .42 |
| When making decisions the principal is aware that his/her assumptions may be incorrect | | | | .40 |
| Eigenvalues | 5.10 | 1.79 | 1.44 | 1.33 |
| % $R^2$ | 30.01 | 10.53 | 8.49 | 7.86 |
| Cronbach α (subscales) | .75 | .78 | .66 | .70 |

*Note*: Principal-axis factoring extraction method. Only values > .40 are shown. *% Cumulative* $R^2$ = 56.91. Cronbach α for whole scale = .85.

**TABLE 4.2** Comparison of Fit Indices of Competing PST Models (N = 100)

| *Model description* | $\chi^2$ | *df* | *Compared with Model 1* $\Delta \chi^2$ | $\chi^2/df$ | *CFI* | *IFI* | *tli* | *RMSEA* |
|---|---|---|---|---|---|---|---|---|
| **Model 1:** Second-order | 139.921 | 110 | | 1.272 | .956 | .958 | .946 | .052 |
| **Model 2:** Four-factor model | 139.141 | 108 | .78 | 1.288 | .954 | .956 | .943 | .054 |
| **Model 3:** Three-factor model | 186.145 | 111 | 46.22** | 1.677 | .890 | .894 | .865 | .083 |
| **Model 4:** One-factor model | 248.175 | 117 | **108.25 | 2.121 | .808 | .813 | .777 | .106 |

*Note:* $\chi^2$ improvement is reflected by a lower value. CFI = comparative fit index; IFI = incremental fit index; TLI = Tucker–Lewis index; RMSEA = root mean square error of approximation.

** $p < .01$.

competencies were the indicators of one second-order latent factor. Model 2 denotes that the 17 observable items load on the four first-order factors (Adopting a multidimensional view; Openness to a variety of opinions; Leading wholes; and Evaluating significance). Model 3 denotes that the 17 items are accounted for by a single first-order factor, PST. Model 4 denotes that the 17 observable items are accounted for by three first-order factors (Evaluating significance; Openness to a variety of opinions; and two factors merged together as one – Leading wholes and Adopting a multidimensional view).

Fit indices (Table 4.2) and the chi-squared differences revealed that the three alternative models were significantly weaker than the second-order model ($\chi$ 2 /df=1.272, CFI = .956, TLI = .946, IFI = .958, RMSEA = .052). This confirmed that PST is a second-order structure (Bollen, 1989). All items had significant loadings on the intended PST characteristics, and all characteristics had significant loadings on the intended second-order latent factor. Figure 4.1 shows the factor loading values of all items in the confirmatory factor analysis, which ranged from .38 to .93.

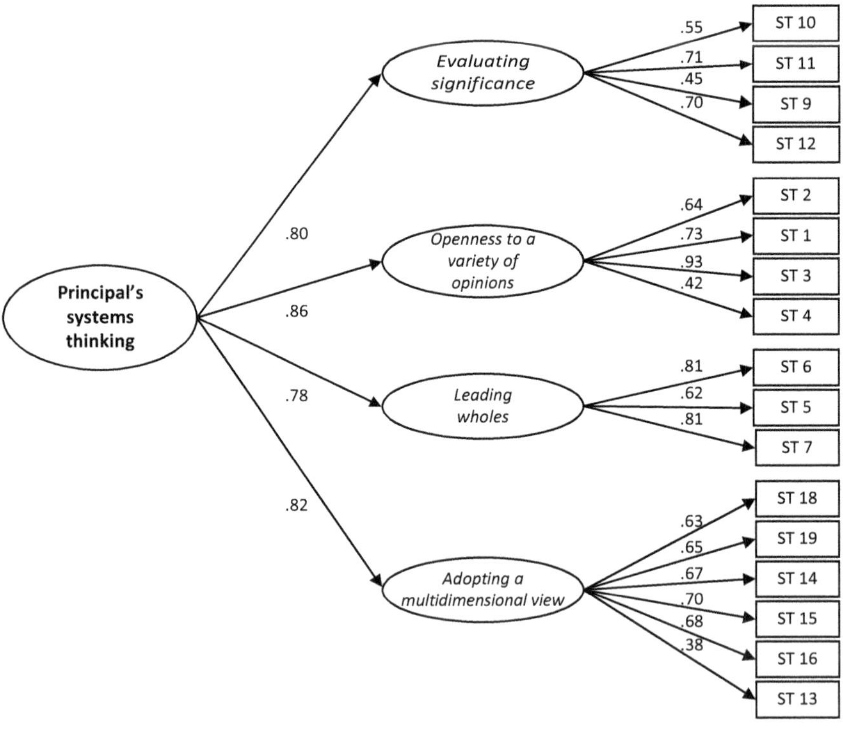


**FIGURE 4.1** Results of the Confirmatory Factor Analysis for 17 Principals' Systems Thinking (PST) Items

*Note*: $\chi^2$ = 139.912; df = 110; CFI = .956; IFI = .958; TLI = .946; RMSEA = .052.

**TABLE 4.3** Range of Factor Loadings, Descriptive Statistics, Alpha Coefficients, and Correlation Analysis for the PST Scale (N = 100)

| *Composite factors* | *No. of items* | *Range of factor loadings* | *M* | *SD* | *Cronbach alpha* | *Correlations* | | |
|---|---|---|---|---|---|---|---|---|
| | | | | | | *2* | *3* | *4* |
| 1 Evaluating significance | 4 | .45 to .71 | 3.42 | .72 | .73 | .53*** | .53*** | .56*** |
| 2 Openness to a variety of opinions | 4 | .42 to .93 | 3.51 | .71 | .83 | 1 | .63*** | .56*** |
| 3 Leading wholes | 3 | .62 to .81 | 3.48 | .80 | .79 | | 1 | .49*** |
| 4 Adopting a multidimensional view | 6 | .38 to .70 | 3.58 | .62 | .78 | | | 1 |

*Note*: Response scale from *Never* (1) to *Always* (5).

*** $p < .001$.

Together, the results of both the exploratory and confirmatory factor analysis support a multidimensional conceptualization of the PST construct, identifying four characteristics. Table 4.3 indicates the range of factor loadings, descriptive statistics, alpha coefficients, and correlation analysis for PST. The results reveal several insights. Overall, in all subscales, the means were close to 4 ("often shown"), indicating high means for all competencies. Second, the results regarding the internal consistency coefficients for the latent factors indicated good results (ranging from .73 to .83). The 17 items of the PST Scale demonstrated a reliability of .90. Finally, the results of the intercorrelations for the PST subscales (see Variables 2–5 on Table 4.4) indicated relatively moderate to high degrees of association between the latent factors ($r = .49–.63$, $p < .001$). These values are appropriate for models that have proposed a priori that the latent factors (characteristics) are interrelated (Brew, Beatty, & Watt, 2004).

**Validation Variables.** To test criterion-related validity, the PST Scale was examined for its correlations with two other well-established constructs: IL and OC. Both the four-factor model and the whole scale were examined for validity criteria. The 100 elementary school mid-level leaders who provided the data for the confirmatory factor analysis responded to both validity criteria surveys: IL and OC.

***Instructional leadership survey***. Validity for the PST Scale was measured using an IL survey completed by the third sample of 100 mid-level leaders. We expected that perceptions of principals' higher ST levels would relate positively to principals' higher IL. For this purpose, we used an abbreviated version of the Principal Instructional Management Rating Scale (PIMRS), originally developed by Hallinger and Murphy (1985) and adapted slightly to accommodate the Israeli context and the elementary school level. The 31-item brief version has shown adequate reliability and validity estimates (Schechter & Qadach, 2016). It measures the following three IL subscales: (a) *Defining the school's mission* (12 items such as "Defines the responsibilities of the team in achieving the school's educational goals"; $\alpha$ = .83); (b) *Managing the instructional program* (10 items such as "Clarifies who is responsible for coordinating the curricula"; $\alpha$ = .86); and (c) *Promoting a positive school-learning climate* (9 items such as "Praises students for high achievements through reinforcements such as prestigious roles or mentioning them in the school paper or site"; $\alpha$ = .81). Mid-level leaders responded on a 5-point Likert scale, ranging from *Never* (1) to *Always* (5). The whole scale's internal reliability was .93.

***Organizational commitment survey***. Validity for the PST Scale was also measured using a survey of the OC experienced by the third sample of 100 mid-level leaders. We expected that principals' higher ST levels would associate positively with their school mid-level leaders' higher OC. For this purpose, we used Mowday et al.'s (1979) 15-item OC scale, which taps the strength of respondents' identification with and involvement in a particular organization, adapted by Somech and Bogler (2002) to suit the current educational setting.

**TABLE 4.4** Descriptive Statistics and Intercorrelations Matrix for Study Variables (N = 100)

| *Variables* | *M* | *SD* | *(1)* | *(2)* | *(3)* | *(4)* | *(5)* | *(6)* | *(7)* | *(8)* | *(9)* | *(10)* | *(11)* | *(12)* | *(13)* |
|---|---|---|---|---|---|---|---|---|---|---|---|---|---|---|---|
| (1) PST: Total | 3.51 | .57 | 1 | .80*** | .82*** | .78*** | .84*** | .64*** | .49*** | .66*** | .57*** | .37*** | .28** | .01 | .39*** |
| (2) PST: *Evaluating significance* | 3.42 | .71 | | 1 | .53*** | .53*** | .56*** | .52*** | .37*** | .58*** | .47*** | .38*** | .24* | .08 | .43*** |
| (3) PST: *Openness to variety of opinions* | 3.51 | .71 | | | 1 | .63** | .56*** | .48*** | .37*** | .49*** | .45*** | .28** | .23* | −.02 | .28** |
| (4) PST: *Leading wholes* | 3.48 | .80 | | | | 1 | .49*** | .47*** | .37*** | .51*** | .40*** | .28** | .22* | −.02 | .32** |
| (5) PST: *Adopting multidimensional view* | 3.58 | .62 | | | | | 1 | .58*** | .48*** | .57*** | .51*** | .27** | .22* | .01 | .26** |
| (6) IL: Total | 3.76 | .62 | | | | | | 1 | .91*** | .87*** | .91*** | .40*** | .30*** | −.11 | .45*** |
| (7) IL: *School mission* | 3.64 | .66 | | | | | | | 1 | .65*** | .78*** | .36*** | .30** | −.09 | .39*** |
| (8) IL: *Instructional program* | 3.94 | .73 | | | | | | | | 1 | .70*** | .42*** | .35*** | .01 | .45*** |
| (9) IL: *Climate* | 3.73 | .70 | | | | | | | | | 1 | .27** | .14 | −.23* | .38*** |
| (10) OC: Total | 3.54 | .49 | | | | | | | | | | 1 | .86*** | .59*** | .91*** |
| (11) OC: *Accepts school goals, values* | 3.30 | .52 | | | | | | | | | | | 1 | .55*** | .63*** |
| (12) OC: *Willing to exert effort* | 3.21 | .66 | | | | | | | | | | | | 1 | .34*** |
| (13) OC: *Maintains membership* | 3.75 | .52 | | | | | | | | | | | | | 1 |

*Note.* IL = principals' instructional leadership. OC = mid-level leaders' organizational commitment.

* $p < .05$; ** $p < .01$; *** $p < .001$.

This measure includes three subscales: (a) *a strong belief in and acceptance of the organization's goals and values* (4 items such as "I find that my values and school values are very similar"; $\alpha = .61$); (b) *a willingness to exert considerable effort on behalf of the organization* (3 items such as "I'm willing to invest a lot of effort, even beyond expectations, to help my school succeed"; $\alpha = .63$); and (c) *a strong desire to maintain membership in the organization* (8 items such as "I'm proud telling others that I work in my school"; $\alpha = .74$). Mid-level leaders responded on a 5-point Likert scale, ranging from *Strongly disagree* (1) to *Strongly agree* (5). The whole scale's internal reliability was $\alpha = .80$.

***Analysis of findings for the two validation constructs.*** Descriptive statistics and intercorrelations of the PST and validity variables collected from this sample ($N = 100$) are presented in Table 4.4. Findings indicated a positive relationship between overall PST and IL ($r = .63$, $p < .001$). Also, significant positive correlations relatively moderate to high emerged between IL and each of the four PST subscales: *Evaluating significance* ($r = .52$, $p < .001$), *Openness to a variety of opinions* ($r = .48$, $p < .001$), *Leading wholes* ($r = .47$, $p < .001$), and *Adopting a multidimensional view* ($r = .58$, $p < .001$). More specifically, as seen in Table 4.4, each of the four PST subscales revealed significant positive correlations at the $p < .001$ level with all three of the IL subscales – *School mission, Instructional program*, and *Climate*.

Furthermore, as predicted, the whole PST Scale was significantly and positively related to the OC survey ($r = .37$, $p < .001$). Also, significant positive correlations emerged between OC and each of the four PST subscales: *Evaluating significance* ($r = .38$, $p < .001$), *Openness to a variety of opinions* ($r = .28$, $p < .01$), *Leading wholes* ($r = .28$, $p < .01$), and *Adopting a multidimensional view* ($r = .27$, $p < .01$). More specifically, as seen in Table 4.4, each of the four PST subscales revealed significant positive correlations with two out of the three OC subscales, *Accepting goals* (at the $p < .05$ level) and *Maintaining membership* (at the $p < .01$ or .001 level), although not with the third OC subscale of *Willingness to exert effort.*

## Discussion of the PST Scale and its Implications

Given that the literature has recommended ST's utilization as an effective approach for managers in general (Brown, 2012; Jolly, 2015; Wilson & Van Haperen, 2015), and for school principals in particular (e.g., Fullan, 2005; Senge et al., 2012), and in light of the lack of available instruments to measure PST, we undertook the development and testing of the multidimensional PST Scale. As seen in this chapter, our exploratory and confirmatory factor analyses provided support for 17 items comprising four PST characteristics, thus providing evidence of construct validity for these characteristics of ST as manifested in elementary school principals: *Adopting a multidimensional view, Openness to a variety of opinions, Leading wholes*, and *Evaluating significance.*

The four characteristics emerging from the current factor analyses of this quantitative scale differ somewhat from the four characteristics that we found in our previous qualitative research (Shaked & Schechter, 2014, 2017), where we conducted content analysis of interview and focus-group data that elicited school principals' articulations about PST. First, *openness to a variety of opinions* was found to be a characteristic rather than a sub-characteristic of PST, as originally proposed in the qualitative studies. A principal who understands that each situation has several aspects and several possible implications seeks to understand the full picture by listening to other points of view. In order to see the whole beyond the parts and to see the parts in the context of the whole, principals must be willing to listen to others and learn from others. Through such openness, principals involve teachers in decision-making and engage in dialogue with those holding educational perceptions that differ from their own. Second, *influencing indirectly*, did not emerge here as a distinguishable characteristic of PST but rather was found to be a sub-characteristic branching out from *adopting a multidimensional view*. Apparently, the use of an indirect approach when dealing with tasks and challenges is not considered a central principal characteristic in the eyes of those mid-level leaders who experience PST in elementary schools. The differences between the findings of this study and the findings of the previous study may be a result of differences in the research method (qualitative vs. quantitative) or in study participants and data collection (self-report by school principals vs. reports of mid-level school leaders on their principals). These differences also may stem from the fact that the present study focused only on primary schools, while participants of the previous study were principals of primary schools, middle schools, and high schools.

Finally, the relationships that we found for PST with principals' IL and with mid-level leaders' OC not only provided further support for the PST Scale's validity but also have important implications. Links between principals' IL and students' achievements have been clearly established through research (Glickman et al., 2014; Robinson, Lloyd, & Rowe, 2008). In addition, teachers' OC was found to be a key component of school success, which positively affects teacher performance and student outcomes (Dee et al., 2006; Park, 2005; Sammons et al., 2007).

Although these findings are encouraging for the school leadership literature, this study's design and context imply some limitations requiring further inquiry before generalizing our outcomes. First, due to our study's cross-sectional design, the data cannot provide direct evidence of causal links between PST and other outcomes like principals' instructional leadership behaviors or mid-level leaders' commitment to their schools. Conceivably, the causal order could be reversed, or causality could be reciprocal. Furthermore, this scale measures principals' behavior that reveals a ST approach to school leadership, but only future research will confirm if such PST tendencies contribute significantly to student achievements or other desired school outcomes. Second, the research was conducted in a specific educational context – elementary schools in Israel – limiting generalizability to other educational and sociocultural contexts and calling for replication in other populations and settings, to possibly substantiate the scale's international validity. Research is already underway to validate this scale for secondary school principals and to examine cross-validation by comparing principals' self-reports with mid-level leaders' ratings in the same schools.

Overall, an instrument measuring PST also holds practical implications, opening up new avenues for research, principal candidate screening, and professional training. The PST Scale can allow for comparisons of possible ST patterns exhibited by principals who differ in experience, gender, education, or type of school (e.g., primary vs. secondary, regular vs. special education). It permits researchers to examine PST's potentially important relations with other variables related to the principal (e.g., distributed leadership, transformational leadership, cognitive complexity), with the school as an organization (e.g., organizational learning, organizational resilience), with the teaching staff (e.g., motivation, self-efficacy), and of course with student outcomes.

Measuring PST may be valuable at various stages throughout principals' careers. The selection process for new principals may be optimized if candidates' capacity for ST is taken into account. In addition to screening prospective principal candidates for their ST skills before appointment, new principals' ST capabilities may be reevaluated after several years on the job. This later reevaluation is required because the development of ST takes time (Shaked & Schechter, 2017), and it unfolds gradually as an outcome of manifold long-term experiences. The reevaluation, accompanied by detailed feedback, may help the growing principal to pinpoint which areas are well-developed and which need further improvement. Moreover, this screening tool may also help superintendents determine which principals have not adequately acquired ST capacities and can even suggest which necessary professional development processes should be tailored to the specific principal to narrow the existing gaps in his or her holistic leadership development. Measuring PST capacities may be timely even after the first few years on the job, because principals need to continuously enhance their skills and consolidate their leadership knowledge. This new scale may thus facilitate the exploration and evaluation of professional training processes aimed at developing PST, from early preservice stages to the mentoring of veteran or experienced principals.

## References

Armstrong, T. (2007). The curriculum superhighway. *Educational Leadership, 64*(8), 16–20.

Arnold, R. D., & Wade, J. P. (2015). A definition of systems thinking: A systems approach. *Procedia Computer Science, 44*, 669–678.

Bogler, R., & Somech, A. (2004). Influence of teacher empowerment on teachers' organizational commitment, professional commitment and organizational citizenship behavior in schools. *Teaching & Teacher Education, 20*(3), 277–289.

Bollen, K. A. (1989). *Structural equations with latent variables*. New York: Wiley.

Brew, C., Beatty, B., & Watt, A. (2004, December). *Measuring students' sense of connectedness with school*. Paper presented at the Australian Association for Research in Education Annual Conference, Melbourne, Australia.

Brown, J. (2012). *Systems thinking strategy: The new way to understand your business and drive performance*. Bloomington, IN: iUniverse.

Cabrera, D., & Cabrera, L. (2015). *Systems thinking made simple: New hope for solving wicked problems*. Ithaca, NY: Odyssean.

Crane, A. (2014). Year coordinators as middle leaders in independent girls' schools: Their role and accountability. *Leading & Managing, 20*(1), 80–92.

Crick, R. D., Barr, S., Green, H., & Pedder, D. (2017). Evaluating the wider outcomes of schools: Complex systems modelling for leadership decisioning. *Educational Management Administration & Leadership, 45*(4), 719–743.

Dee, J. R., Henkin, A. B., & Singleton, C. A. (2006). Organizational commitment of teachers in urban schools: Examining the effects of team structures. *Urban Education, 41*(6), 603–627.

Elm, J. P., & Goldenson, D. R. (2012). *The business case for systems engineering study: Results of the systems engineering effectiveness survey*. Pittsburgh, PA: Carnegie Mellon University.

Fleming, P., & Amesbury, M. (2012). *The art of middle management: A guide to effective subject, year and team leadership*. New York: Routledge.

Frank, M. (2012). Engineering systems thinking: Cognitive competencies of successful systems engineers. *Procedia Computer Science, 8*, 273–278.

Fullan, M. (2005). *Leadership and sustainability: System thinkers in action*. Thousand Oaks, CA: Corwin.

Fullan, M. (2014). *The principal: Three keys to maximizing impact*. San Francisco, CA: Jossey-Bass.

Gharajedaghi, J. (2011). *Systems thinking, managing chaos and complexity: A platform for designing business architecture* (3rd ed.). Burlington, MA: Morgan Kaufmann.

Glickman, C. D., Gordon, S. P., & Ross-Gordon, J. M. (2014). *Supervision and instructional leadership: A developmental approach* (9th ed.). London, UK: Pearson.

Gurr, D., & Drysdale, L. (2013). Middle-level secondary school leaders: Potential, constraints and implications for leadership preparation and development. *Journal of Educational Administration, 51*(1), 55–71.

Hair, J. F., Anderson, R. E., Tatham, R. L., & Black, W. C. (1998). *Multivariate data analysis* (5th ed.). Upper Saddle River, NJ: Prentice Hall.

Hallinger, P. (2011). A review of three decades of doctoral studies using the Principal Instructional Management Rating Scale: A lens on methodological progress in educational leadership. *Educational Administration Quarterly, 47*(2), 271–306.

Hallinger, P. & Murphy, P. (1985). Assessing the instructional management behavior principals. *Elementary School Journal, 86*(2), 217–247.

Hargreaves, A., & Braun, H. (2013). *Data-driven improvement and accountability*. Boulder, CO: National Education Policy Center.

Hieronymi, A. (2013). Understanding systems science: A visual and integrative approach. *Systems Research and Behavioral Science, 30*(5), 580–595.

Hitchins, D. K. (2007). *Systems engineering: A 21st century systems methodology*. Chichester, UK: John Wiley.

Holmes, B. J., Finegood, D. T., Riley, B. L., & Best, A. (2012). Systems thinking in dissemination and implementation research. In R. C. Brownson, G. A. Colditz, &

E. K. Proctor (Eds.), *Dissemination and implementation research in health: Translating science to practice* (pp. 175–191). New York: Oxford.
Jackson, M. C. (2010). Reflections on the development and contribution of critical systems thinking and practice. *Systems Research and Behavioral Science, 27*(2), 133–139.
Jolly, R. (2015). *Systems thinking for business: Capitalize on structures hidden in plain sight.* Portland, OR: Systems Solutions.
Jöreskog, K., & Sörbom, D. (1993). *LISREL 8: Structural equation modeling with the SIMPLIS command language.* Chicago, IL: Scientific Software International.
Kaiser, H. F. (1958). The varimax criterion for analytic rotation in factor analysis. *Psychometrics, 23,* 187–201.
Kline, P. (1998). *The new psychometrics: Science, psychology and measurement.* Florence, KY: Taylor and Francis/Routledge.
Leischow, S. J., Best, A., Trochim, W. M., Clark, P. I., Gallagher, R. S., Marcus, S. E., & Matthews, E. (2008). Systems thinking to improve the public's health. *American Journal of Preventive Medicine, 35*(2S), S196–S203.
Louis, K. S., Leithwood, K., Wahlstrom, K. L., & Anderson, S. E. (2010). *Learning from leadership: Investigating the links to improved student learning.* New York: Wallace Foundation.
May, H., & Supovitz, J. A. (2011). The scope of principal efforts to improve instruction. *Educational Administration Quarterly, 47*(2), 332–352.
Mowday, R. R., Steers, R. M., & Porter, L. W. (1979). The measurement of organizational commitment. *Journal of Vocational Behavior, 14,* 224–247.
Murphy, J., Neumerski, C. M., Goldring, E., Grissom, J., & Porter, A. (2016). Bottling fog? The quest for instructional management. *Cambridge Journal of Education, 46*(4), 455–471.
Neumerski, C. M. (2012). Rethinking instructional leadership, a review: What do we know about principal, teacher, and coach instructional leadership, and where should we go from here? *Educational Administration Quarterly, 49*(2), 310–347.
Ng, F. S. D. (2015). Leadership learning for complex organizations. *Cogent Education, 2*(1), Article 1017312.
Park, I. (2005). Teacher commitment and its effects on student achievement in American high schools. *Educational Research and Evaluation, 11*(5), 461–485.
Richmond, B. (1994). System dynamics/systems thinking: Let's just get on with it. *System Dynamics Review, 10*(2–3), 135–157.
Richmond, B. (2000). *The "thinking" in systems thinking: Seven essential skills.* Waltham, MA: Pegasus.
Rigby, J. G. (2014). Three logics of instructional leadership. *Educational Administration Quarterly, 50*(4), 610–644.
Robinson, V. M. J., Lloyd, C., & Rowe, K. (2008). The impact of leadership on student outcomes: An analysis of the differential effects of leadership types. *Educational Administration Quarterly, 44*(5), 564–588.
Saiti, A. (2015). Conflicts in schools, conflict management styles and the role of the school leader: A study of Greek primary school educators. *Educational Management Administration & Leadership, 43*(4), 582–609.
Sammons, P., Day, C., Kington, A., Gu, Q., Stobart, G., & Smees, R. (2007). Exploring variations in teachers' work, lives and their effects on pupils: Key findings and implications from a longitudinal mixed-method study. *British Educational Research Journal, 33*(5), 681–701.

Schechter, C., & Qadach, M. (2016). Principals' learning mechanisms: Exploring an emerging construct. *Leadership and Policy in School, 15*(2), 141–167.

Senge, P. (1990). *The fifth discipline: The art and practice of the learning organization.* New York: Currency, Doubleday.

Senge, P. (2006). *The fifth discipline: The art and practice of the learning organization* (2nd ed.). New York: Currency, Doubleday.

Senge, P. M., Cambron-McCabe, N., Lucas, T., Smith, B., Dutton, J., & Kleiner, A. (2012). *Schools that learn: A fifth discipline fieldbook for educators, parents and everyone who cares about education.* New York: Crown.

Shaked, H., & Schechter, C. (2014). Systems school leadership: Exploring an emerging construct. *Journal of Educational Administration, 52*(6), 792–811.

Shaked, H., & Schechter, C. (2017). *Systems thinking for school leaders: Holistic leadership for excellence in education.* Cham, Switzerland: Springer.

Somech, A., & Bogler, R. (2002). Antecedents and consequences of teacher organizational and professional commitment. *Educational Administration Quarterly, 38*(4), 555–577.

Sterman, J. D. (2000). *Business dynamics: Systems thinking and modeling for a complex world.* Boston, MA: McGraw-Hill.

Steiger, J. H. (1990). Structural model evaluation and modification: An interval estimation approach. *Multivariate Behavioral Research, 25*, 173–180.

Stevens, J. (1996). *Applied multivariate statistics for the social sciences* (3rd ed.). Mahwah, NJ: Lawrence Erlbaum.

Stronge, J. H., Richard, H. B., & Catano, N. (2008). *Qualities of effective principals.* Alexandria, VA: Association for Supervision and Curriculum Development.

Supovitz, J., Sirinides, P., & May, H. (2010). How principals and peers influence teaching and learning. *Educational Administration Quarterly, 46*(1), 31–56.

Tejeda, J., & Ferreira, S. (2014). Applying systems thinking to analyze wind energy sustainability. *Procedia Computer Science, 28*, 213–220.

Thorpe, A., & Bennett-Powell, G. (2014). The perceptions of secondary school middle leaders regarding their needs following a middle leadership development programme. *Management in Education, 28*(2), 52–57.

Walker, J., & Slear, S. (2011). The impact of principal leadership behaviors on the efficacy of new and experienced middle school teachers. *National Association of Secondary School Principals Bulletin, 95*(1), 46–64.

Wilson, B., & Van Haperen, K. (2015). *Soft systems thinking, methodology and the management of change.* London: Palgrave.

**PART II**

# Systems Thinking at the District/Regional Level

# 5

# USING HIERARCHICAL GROWTH MODELING TO PROMOTE DISTRICT SYSTEMATIC IMPROVEMENT IN OHIO AND TEXAS

*Alex J. Bowers, Xinyu Ni, and Jennifer Esswein*

Over the past 30 years, the research and practice literature on the effects of school districts on student achievement has indicated that school district personnel practices can have a strong effect on individual and overall student, school and district achievement (Bryk, 2010; Elmore & Burney, 1997; Honig, 2003, 2008; Leithwood, 2010; Purkey & Smith, 1985). Additionally, as the United States national educational policy changes from rigid sanction rules and high pressure under No Child Left Behind (NCLB) to a more flexible and local capacity framework of the Every Student Succeeds Act (ESSA), it is important for researchers, practitioners, and policy-makers to consider school districts as the leverage point for bringing multiple financial and human capital components of the district to bear on district-level improvement, rather than an emphasis on school-by-school change as districts in the United States are made up of systems of schools nested within the district community (Daly & Finnigan, 2016). Indeed, while some school districts struggle to find success, others outperform peer districts that have similar context and demographic effects over long periods of time. However, previous research and practice has encountered obstacles in sampling and site-selection in what has been called District Effectiveness Research (DER) (Bowers, 2010). According to Trujillo (2013, 2016), the majority of studies in DER rely on convenience samples or districts with extremely high or low performance without considering district demographic characteristics, which makes it difficult to assess the external validity and generalizability of these research results.

Thus, in addressing these issues, the purpose of this study is to apply hierarchical linear growth modeling to all school districts in the states of Ohio and Texas in the United States to identify districts that significantly outperform or underperform their contexts, demographics, and resources in comparison to the rest of the system of districts across the population of a state. The study builds on the previous literature on system-level research at the district level to identify unusually effective school districts over time in three main ways. First, we applied the site-selection hierarchical longitudinal model (Bowers, 2008, 2010, 2015) in two different states, Ohio and Texas, each with very different educational histories and challenges, and we successfully identify outperforming districts, demonstrating the generalizability and practicability of this modeling framework in DER. Second, we propose the system-level district effectiveness analysis using data from every school district in a state (the system-level) which can then be used to bring together personnel from top districts to discuss what is idiosyncratic versus generalizable in their work in their districts and states in order to help other districts across the system improve. Third, we discuss how the outperforming districts identified from the model compare to the "excellent" districts as graded by each state accountability system to provide further suggestions on state evaluation systems.

Throughout this study we consider school districts as the central actor in a systems perspective, in which we draw on a central aspect of much of the recent research on the school district as being made up of not only a system of schools, but a system of central office personnel and school building leaders in addition to and in collaboration with the superintendent and elected school boards (Daly & Finnegan, 2016; Honig, 2003, 2008; Honig & Rainey, 2015). These researchers have argued for districts research to go beyond considering "the district" as a monolithic actor, and instead consider the work of the many different people throughout a school district that make it work. In the present study we expand on this conception, and examine districts within the broader system of school districts across two states, considering every district within the policy region, over long periods of time. From a systems-thinking perspective, this helps move the research from the previous "snapshot" single district or timepoint studies of the past research (Bowers, 2010, 2015; Trujillo, 2013, 2016) towards "a framework for seeing interrelationships rather than things, for seeing patterns of change rather than static snapshots" (Senge, 1990, p. 68) and "seeing beyond what appear to be isolated and independent incidents to deeper patterns" (O'Connor & McDermott, 1997, p. 7).

## Background and Literature Review

In the United States, public schools are nested within school districts, which makes the school district a unique unit for analysis in research and practice. A continual question within the research on school districts has looked to find unusually effective districts, known as district effectiveness research (DER), originating in the 1980s. In the early district effectiveness research (DER), researchers mainly focused on central office practices that were linked with school and district success as well as what characteristics were associated with successful school districts.

For instance, Purkey and Smith (1985) argued that school districts could promote successful school reforms through several practices such as setting clear goals at the school level and providing coherent and organized staff development. Leithwood (2010) examined 31 previous studies on district effectiveness and summarized 10 district characteristics that are mostly effective in promoting students' educational outcomes including setting instructional goals, promoting instructional improvement, and aligning multiple resources such as personnel policies and financial allocations. The majority of the research in DER indicates that school districts have a strong impact in promoting students' learning, advancing teaching quality, focusing on sustained school-level professional development and closing achievement gaps (Bowers, 2008; Bowers & Lee, 2013; Bryk, 2010; Elmore & Burney, 1999; Firestone, Mangin, Martinez, & Polovsky, 2005; Honig, 2003, 2008; Levin, Datnow, & Carrier, 2012; Opfer, Henry, & Mashburn, 2008). However, in a recent study (Chingos, Whitehurst, & Gallaher, 2015) analyzing data from Florida and North Carolina fourth and fifth grades with a cross-sectional hierarchical linear model, the authors found that the district effect explained only a small amount of the variation in student achievement – with one standard deviation increase in district effectiveness associated with 0.07–0.14 standard deviation increase in student achievement in reading or mathematics for single years.

Nevertheless, across the DER literature, authors note that the study of school districts is a complicated concept, as an increasing amount of researchers agree that districts should not be considered as a monolithic concept of "the district," which oftentimes focuses on superintendent behavior, but rather instead, the district is an organizational and systems concept made up of multiple demographic, sociopolitical and technical elements, which vary in size, characteristics, conditions and context (Daly & Finnigan, 2012, 2016; Honig & Rainey, 2015). Thus, building on this previous literature, we argue here that the DER literature should model district effects using a more dynamic and systems-level perspective, rather than focus on single-year or "snapshot" research models (Bowers, 2010, 2015).

First, there is a need to measure the effectiveness of districts through more dynamic and systematic ways, especially as the DER literature has noted that the definition of effectiveness ranges broadly from standardized test score performance, to student growth in test scores, to other outcomes, such as high school graduation and college-going, to name just a few (Bowers, 2010, 2015). As a measure of collaborative efforts from the central office within school districts, the measurement of "district effectiveness" should reflect an index of multiple aspects of district school performance, such as student overall performance on different subjects from various grades as well as teacher effectiveness. However, as noted in the previous research (Bowers 2010; Trujillo, 2013), a majority of studies have relied on just English/reading or mathematics standardized state test scores to conceptualize "effectiveness." This thus unnecessarily overly restricts the definition of effectiveness.

Second, district effects happen over time, so it is important to conduct longitudinal studies to explore the improvement or change of district effects over multiple years. Specifically, as districts in the U.S. have come under increasing accountability demands over the past 30 years (Ni, Bowers, & Esswein, 2016), effective districts could serve as a key link between state-context policies and school day-to-day practices through developing "coherent instructional strategies" (Thompson, Sykes, & Skrla, 2008) in which they bring together the multiple financial and human capital components of the district in service to promote overall systematic changes (Skrla, McKenzie, Scheurich, & Dickerson, 2011). But such organizational improvements would happen continuously and slowly (Smylie, 2009), which cannot be detected through studies that rely on data from a single point in time. Thus, continuous examination through longitudinal studies is necessary to explore and reflect on the overall systematic changes within districts.

Third, researchers need to develop a systematic and comprehensive method to identify or define "effective districts" for in-depth follow-up qualitative studies by taking district demographic characteristics into account. As noted above, the goal of DER has historically been to identify unusually effective school districts, and then work to study those districts in comparison to districts at the average through in-depth qualitative studies to identify the practices and innovations that may lead to effectiveness at the district level in a state. However, as an organization composed of various resources and processes in the context of the community in which it is situated, school performance is historically strongly correlated with community demographics, such as poverty or wealth, as well as the proportion of students from historically disadvantaged backgrounds. Simply identifying highly effective districts from the ranking of a certain measurement, such as student academic achievement, without considering these types of unique district characteristics will identify only districts that are comparatively well-off. This is problematic when attempting to find the most effective districts, as organizations should not be penalized for variables that are outside of their control, such as student and community socio-economic status. This issue can be addressed through including all districts in a state in the analysis and statistically controlling for these demographic and context effects (Bowers, 2008, 2010, 2015; Leithwood, 2010).

In a series of longitudinal studies, Bowers (2008, 2010, 2015) detailed a site-selection model through applying a hierarchical linear growth modeling approach to all school districts in Ohio to identify districts that significantly outperform or underperform their background and demographics, while including sociopolitical and technical resources. These studies nested time within districts, and then modeled both the performance of each district in the first year of the model, as well as growth over time, examining which districts outperformed in overall performance as well as growth over the period, while controlling for background and context variables. In an attempt to examine the systematic changes for all districts in Ohio and address previous critiques, the statistical models relied on seven consecutive years of data, including demographic variables for each district, and a composite measure of district effectiveness – the Performance Index (PI) score – which is an aggregated measure of students' performance on state standard tests for reading, mathematics, science and social studies across grades 3 through 12. The model successfully identified groups of outperforming school districts that achieved significant performance growth in comparison to the entire population of school districts in Ohio over an extended period of time. These districts were then identified by name, and deemed as having significantly outperformed the average district with similar demographic and context variables, and were then nominated for in-depth qualitative analysis with the purpose of comparing practices in outperforming school districts with the norm or the underperforming school districts. This site-selection strategy provides a potential way to examine systematic and dynamic district changes, generates knowledge that may help inform decision-making, and improves external validity for research studies in DER (Trujillo, 2013). Also, given that the model was conducted on two overlapping but different sets of years in the two studies (Bowers, 2010, 2015), this showed that the statistical model works successfully with different time periods within the single state of Ohio.

Nevertheless, Bowers' (2010) and (2015) studies are limited in three major aspects. First, the site-selection model has only been applied to Ohio to date and there is no evidence indicating whether the model will work in states having more districts and larger variance in demographic characteristics than Ohio. In the present study, we address this issue through replicating the analysis in Texas, which is a much larger state than Ohio in both population as well as number of districts and schools (over 1,000 districts in Texas compared to about 600 in Ohio), and Texas has a different set of state assessments than Ohio, their own set of education policies and laws and different state governments. Additionally, Texas is demographically and geographically a very different context from Ohio, as Ohio is a mid-western state in the industrial rust-belt of the US, while Texas is a large southern border state with a large and growing Hispanic/Latino population and a growing industrial base. Second, although previous studies have included average teacher salary in the model, more detailed financial expenditures haven't been considered, such as direct expenditures on instructional services by the district. Previous qualitative studies have indicated that central office personnel in effective districts are able to align multiple funding streams and financial components with the instructional goals of the district to address instructional and pedagogical improvement across all schools in the system (Bowers, 2008; Elmore & Burney, 1997; Skrla et al., 2011; Thompson et al., 2008). Therefore, more savvy financial indicators in addition to teacher average salary are needed, as if a district is outperforming context and demographic variables as well as financial variables of the districts at comparable averages in a state, then an outperforming district in the model could be considered as effective, equitable and efficient. Third, the model successfully identifies groups of outperforming districts compared to the majority of districts at the norm. This is important as past DER studies provide little in the way of comparison to the three different groups of districts of outperforming, at the norm (the majority), and underperforming. However, to date this modeling strategy has provided little in the way of helping to compare the model-identified outperforming districts to the "excellent" districts graded by the state accountability system. These comparisons are important, as the Bowers (2010, 2015) modeling framework is not intended to replace accountability systems, but rather to inform the study of districts throughout a state policy region. There is an interesting conversation to be had at the district and state level when these two systems agree or do not agree, and why.

### *Framework of the Study*

The purpose of this study is to address the above issues and apply recent district effectiveness research (DER) site identification models across states, using hierarchical linear growth models in Ohio and Texas, including a broader set of time-varying demographic and financial covariates over an extensive period of time in both states. We identify groups of districts from the entire population of a state that are significantly outperforming peer districts. The goal is to further study outperformers as identified by the analyses in order to determine generalizable improvement approaches. For example, central office administrators from the outperforming districts could be brought together in collaboration with each state's department of education to provide professional development opportunities that focus on promoting state-level cross-district information sharing and improvement cycles. The reason we select these two states is to first replicate the previous results in Ohio with adding the additional finance variables noted above, and then second to apply to a very different state context, Texas, to examine the extent to which the modeling process will replicate to a very different state context, as far as geography, number of districts, school sizes, demographics, and achievement are concerned.

In the present study, our research questions are: 1) To what extent can outperforming or underperforming school districts in Ohio and Texas be identified by the model based on the long-term trends of multiple indicators? 2) To what extent is this identification system consistent with each state's current rating system? 3) To what extent do the district financial variables play a role in the growth trajectories of school districts for both Ohio and Texas?

## Method

### *Sample*

This study is a secondary data analysis of the publicly available data from the Ohio Department of Education (ODE) and Texas Education Agency (TEA).

Specifically, we focused on administrative panel data for all school districts in Ohio (*n*=608) during the academic years of 2005–2006 to 2011–2012 and for all districts in Texas (*n*=1028) during the academic years of 2004–2005 to 2011–2012. Additionally, for both states, we merged the state data with publicly accessible data from the National Center for Education Statistics (NCES) Common Core of Data (CCD) and the National Financial Survey to complement the state data with district financial, demographic and context variables.

## Variables

The dependent variable in the Ohio model is the district Performance Index (PI) Score, which captures the yearly performance of students across multiple grades for all tested subjects including mathematics, reading, social studies and science over a seven-year time span (2006–2012). According to the calculation algorithm provided by ODE (2014) and Bowers (2010, 2015), the PI score is calculated as a weighted average of the percentages of student performance in advanced, basic and below basic categories for students in grades 3, 8 and 10 for all tested subjects, including reading, mathematics, science and social studies, with higher performance categories receiving higher weights after accounting for testing rates.

Because Texas did not report a single performance indicator for districts over this time period, the dependent variable in the Texas model is constructed from the publicly available Texas Education Agency (TEA) dataset. The TEA dataset includes data on each district's scores on the standardized tests of Texas Assessment of Knowledge and Skills (TAKS), which measures students' abilities in reading, mathematics, science, social studies and writing across grades 3 to 11 for the years 2002–2012. After examining the missingness pattern of the whole dataset, we included in the final analysis the data from all non-charter public school districts in the state from 2004 to 2011 (n=1028). Lacking a single index measuring continuous performance and various aspects of school districts in Texas, we first calculated a PI score for each district in Texas following the Ohio PI score calculation metrics based on a recent report that summarizes similar PI metrics for all states in the U.S. (Ni et al., 2016).

## Independent Variables

For the independent variables in both models, we followed previous literature on school and district effects (Bowers, 2010, 2015), which included *Percentage of Disadvantaged Students, Percentage of Asian Students, Percentage of African American Students, Percentage of Hispanic Students*, and *Percentage of Disabled Students. Percentage of High Mobility Students* defined by ODE (2014) as the percentage of students who had attended schools in the district for two or fewer academic years. We then created three dichotomous variables to indicate urbanicity, (i.e., Urban, Small Town, and Rural, with Suburban as reference group) according to the local code variable listed by NCES CCD for each district (NCES, n.d.). The variables *Percentage of Student Attendance, Average Years of Teacher Experience, Student-Teacher Ratio*, and *Average Teacher Salary*, and *School-level PI variance* were included in the analysis as in the past studies (Condron & Roscigno, 2003; Ferguson, 1991; Finn, Gerber, & Boyd-Zaharias, 2005; Hanushek, 1989; Verstegen & King, 1998). In addition, following the recommendations of previous research on categorizing district enrollment (Bowers & Urick, 2011), we created the categories for enrollment of *Small* (1–1,299), *Large* (2400–9999) and *Extra Large* (10,000 students and more), with medium enrollment as the reference group (1,300–2,399) in the Ohio model. However, in the Texas model, we categorized the enrollment data into five subgroups by adding an extra small category so as to better characterize the districts that have yearly enrollment of less than 650.

**TABLE 5.1** Descriptive of Ohio District Variables 2005-06 – 2011-12

| *Variable* | *Mean* | *(SD)* | *Min* | *Max* |
|---|---|---|---|---|
| Performance Index (PI) Score | 97.05 | (6.27) | 69.80 | 112.40 |
| % Disadvantaged students | 0.34 | (0.02) | 0 | 1.00 |
| % Asian students | 0.01 | (0.016) | 0 | 0.17 |
| % African American students | 0.06 | (0.14) | 0 | 0.99 |
| % Hispanic students | 0.02 | (0.03) | 0 | 0.43 |
| % High Mobility students | 0.39 | (0.11) | 0.15 | 1 |
| Small enrollment (vs. medium) | 0.35 | (0.48) | 0 | 1.00 |
| Large enrollment (vs. medium) | 0.30 | (0.46) | 0 | 1.00 |
| Extra Large enrollment (vs. medium) | 0.02 | (0.15) | 0 | 1.00 |
| Urban (vs. suburban) | 0.04 | (0.18) | 0 | 1 |
| Small Town (vs. suburban) | 0.12 | (0.33) | 0 | 1 |
| Rural (vs. suburban) | 0.47 | (0.50) | 0 | 1.00 |
| % Student attendance | 0.95 | (0.016) | 0.89 | 0.98 |
| School-level PI score variance | 4.11 | (3.04) | 0.00 | 36.92 |
| Avg. teacher years experience | 14.85 | (3.27) | 4.00 | 34.00 |
| Student – teacher ratio | 15.94 | (1.97) | 0.87 | 49.73 |
| Avg. teacher salary ($) | 51560.79 | (7118.13) | 30,810 | 81851 |
| Instruction expenditure per pupil ($) | 5416.93 | (897.75) | 3719.96 | 10808.95 |
| General Administration expenditure per pupil ($) | 264.22 | (122.68) | 44.94 | 2635.61 |
| School Administration expenditure per pupil ($) | 540.12 | (127.51) | 210.32 | 1495.32 |
| Instruction Staff Support Service expenditure per pupil ($) | 456.56 | (236.90) | 11.71 | 1819.45 |
| Capital Outlay expenditure per pupil ($) | 1445.90 | (3354.84) | 0 | 52896.96 |
| n | 608 | | | |
| Number of years of data | 7 | | | |

The model in the analysis extends the previous research (Bowers, 2010, 2015) by examining the effects of the financial variables on the average Ohio and Texas district growth in PI score through the seven-year timespan. Following the previous literature (Card & Payne, 2002; Okpala, Okpala, & Smith, 2001; Payne & Biddle, 1999), we included five financial variables in the current study from the National Public Education Financial Survey, which are *Instruction expenditure per pupil, General Administration expenditure per pupil, School Administration per pupil, Instruction Staff Support Service expenditure per pupil,* and *Capital Outlay – Construction expenditure per pupil.* Tables 5.1 and Table 5.2 summarize the descriptive statistics for the variables in the Ohio and Texas models.

**TABLE 5.2** Descriptive of Texas District Variables 2004-05 – 2010-11

| *Variable* | *Mean* | *(SD)* | *Min* | *Max* |
|---|---|---|---|---|
| Performance Index (PI) Score | 92.95 | (7.05) | 55.70 | 113.60 |
| % Disadvantaged students | 0.54 | (0.18) | 0 | 1.00 |
| % Asian students | 0.09 | (0.02) | 0 | 0.26 |
| % African American students | 0.08 | (0.12) | 0 | 0.86 |
| % Hispanic students | 0.33 | (0.27) | 0 | 1.00 |
| Extra Small enrollment (vs. medium) | 0.39 | (0.49) | 0 | 1.00 |
| Small enrollment (vs. medium) | 0.20 | (0.40) | 0 | 1.00 |
| Large enrollment (vs. medium) | 0.19 | (0.39) | 0 | 1.00 |
| Extra Large enrollment (vs. medium) | 0.09 | (0.29) | 0 | 1.00 |
| Urban (vs. suburban) | 0.07 | (0.25) | 0 | 1.00 |
| Small Town (vs. suburban) | 0.15 | (0.36) | 0 | 1.00 |
| Rural (vs. suburban) | 0.60 | (0.49) | 0 | 1.00 |
| % Student attendance | 0.95 | (0.01) | 0.92 | 0.99 |
| School-level PI score variance | 6.06 | (0.49) | 0.002 | 56.05 |
| Avg. teacher years experience | 12.45 | (2.38) | 0 | 29.50 |
| Avg. teacher salary ($) | 41468.14 | (4355.97) | 29209.16 | 74661.05 |
| Student-Teacher ratio | 0.09 | (0.02) | 0.03 | 0.30 |
| Instruction expenditure per pupil ($) | 40512.78 | (201670.40) | 5.25 | 6015561.29 |
| General Administration expenditure per pupil ($) | 1081.85 | (3061.45) | 0.71 | 79032.43 |
| School Administration expenditure per pupil ($) | 3802.93 | (2222.80) | 0 | 687806.45 |
| Instruction Staff Support Service expenditure per pupil ($) | 3221.90 | (19278.99) | 0 | 623826.92 |
| Capital Outlay expenditure per pupil ($) | 13376.80 | (65445.79) | 0 | 2065129.03 |
| n | 1028 | | | |
| Number of years of data | 7 | | | |

## *Analysis*

Our analysis strategy takes into account the recent critiques of Value Added Models (VAMs) (Hallinger, Heck, & Murphy, 2014; Murphy, Hallinger, & Heck, 2013) through using a hierarchical linear growth model to examine longitudinal district performance trends (Bowers & White, 2014; Hox, 2010) in both Texas and Ohio, controlling for performance in the first year and then estimating the change over time as the slopes. We used SPSS 22 for all analyses. To correct for normality, we used a square root transformation for the ethnicity and financial variables. Also, we standardized all of the continuous variables into z-scores to grand mean center. Following the recommendations for specifying the covariance structure for repeated measures (Heck, Thomas, & Tabata, 2013), a diagonal error covariance matrix (DIAG) was specified at level 1 while an unstructured error covariance matrix (UN) was specified at level 2. The full model analyzed in the analysis can be written as:

Level 1:

$$PI_{ti} = \pi_{0i} + \pi_{1i}\,\mathrm{Year}_{ti} + e_{ti}$$

Level 2:

$$\pi_{0i} = \beta_{00} + \beta_{01}\,X_{ti} + \ldots + u_{0i}$$

$$\pi_{1i} = \beta_{10} + \beta_{11}\,X_{ti} + \ldots + u_{1i}$$

in which:

$PI_{ti}$ = district PI score for district $i$ in time $t$

$X_{ti}$ = time – varying independent variables for each district $i$ in time $t$

$Year_{ti}$ = year $t$ of each district $i$

$\pi_{0i}$ = intercepts varying randomly for each district $i$

$\pi_{1i}$ = slopes varying randomly for each district $i$

$e_{0i}$ = error in predicting PI for district $i$ in time $t$

$u_{0i}$ = level 2 random components in predicting intercepts for district $i$

$u_{1i}$ = level 2 random components in predicting slopes for district $i$

Finally, we created two sets of plots to visualize the results. First, following Bowers (2010, 2015), we plotted the model predicted gain versus the actual gain for both Ohio and Texas models. We calculated the model predicted PI scores for each district in the dataset by subtracting the first year PI score from the last year predicted PI score, and the same was done for actual PI score gain. The two gain score calculations were then plotted against each other to generate the final figure. Second, we created a quadrant plot for each state to compare the study findings to the state accountability ratings. The y-axis in the quadrant plot indicates the model predicted results, which we defined as PI Gain Index. We calculated PI Gain Index by subtracting the model predicted gain from the actual PI score gain. The x-axis is the school letter rating as defined by each state. In the Ohio model, the x-axis is the percentage of the standards met, scaling from 0 to 100; in the Texas model, the x-axis is the 4-year High School Completion Rate.

## Results

In this section, we first present the overall PI score trajectories for all districts in both Ohio and Texas to provide a visualization of the change-over-time for all districts in each state. Second, we present the fit of the model to the data in both Ohio and Texas, and examine the extent to which different context, demographic and financial variables affect the model. Third, we present the comparison of predicted PI score gains versus actual PI score gains and discuss the "outperforming" districts identified by the models in both states. Finally, we visualize the outperforming districts identified from the model compared to the "excellent" districts graded by the state accountability system to provide further considerations for state evaluation systems in both states.

### *Examining Ohio and Texas School District PI Score Trajectories*

Figure 5.1 plots the seven-year longitudinal trajectories in PI score for each of the 608 districts in Ohio, from the academic years of 2005–2006 to 2011–2012, as each line represents one district. Figure 5.2 plots the seven-year trajectories in PI score for each of the 1,028 districts in Texas. The range of Performance Index score was from 0 to 120, however, there were no districts with a PI score under 50. Overall for each state, Figures 5.1 and 5.2 show that while there is a slight increasing trajectory over the seven years, and there are a few districts at the lower end of the scale with more variance year-over-year, in general the vast majority of school districts perform similar to the previous year and mirror the overall slight trajectory increase for the entire state.

### *Hierarchical Linear Growth Model of Ohio and Texas PI Change*

The unconditional null hierarchical linear growth model with time included for Ohio had a high interclass correlation coefficient (ICC) of 75.56%, indicating

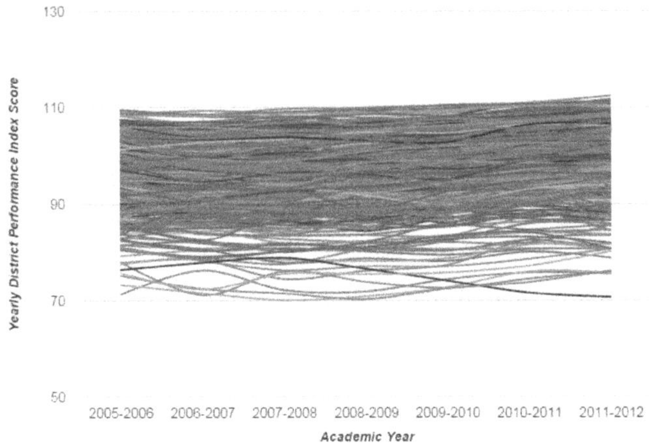


**FIGURE 5.1** Seven-Year Trends in 608 Ohio District PI Scores

Seven years of consecutive PI score data for the full population of school districts in the state of Ohio are plotted. Each line represents the trend of one of the 608 individual district seven-year history of PI score change. Districts trends are plotted from 50 to 120, since no districts scored below 50 PI score at any time.

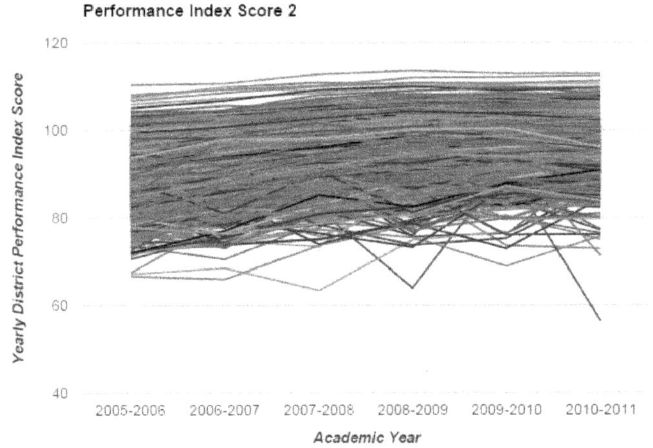


**FIGURE 5.2** Seven-Year Trends in 1028 Texas District PI Scores

Seven years of consecutive PI score data for the full population of school districts in the state of Texas are plotted. Each line represents the trend of one of the 1,028 individual district seven-year history of PI score change. Districts trends are plotted from 40 to 120, since no districts scored below 40 PI score at any time.

that over half of the variance in PI score is between districts. This high between district variance indicates that the growth over time within districts was fairly low, with more variance between districts. This mirrors Figure 5.1 in which one can see the variance up and down the scale between districts, with only slight changes within districts over time (left to right). Table 5.3 provides the results for the full seven-year hierarchical linear growth model. All variables are rescaled (z-scored or categorized) such that all coefficients can be interpreted for the "average" district, a district that would be at the averages for all continuous variables and the reference groups for the categorical variables (medium enrollment and suburban). The unconditional null hierarchical growth model with time included for Texas has an ICC of 50.49%, indicating more than half the variance in PI score can be explained between districts. This is evidenced in Figure 5.2 as there is much more variance in change over time as many districts rise over time while others show more of a "sawtooth" effect over time, especially for the lower performing districts. Table 5.3 provides the results for the full seven-year hierarchical linear growth model. All variables are z-scored or categorized so that all coefficients for all continuous variables can be interpreted as the average effect for all the 1,028 districts.

Parameter estimates for each model replicate and extend the previous studies in Ohio (Bowers, 2010, 2015). Hierarchical linear growth models have two components, the intercept part of the model which controls for prior achievement and context effects, which is analogous to a single cross-sectional regression for the first year of data included in the model, and then the slope component of the model, which examines the effect of each variable on the growth over time in PI score for each state. For both Ohio and Texas, the demographic and context variables generally mirror the previous studies, both in effect size and significance in both the intercept and slope components of the model. This demonstrates that the modeling framework generalizes well from Ohio to the Texas context.

Next, following the recommendations from the previous Ohio studies (Bowers, 2010, 2015), we subtracted the 2012 Actual PI score for each district from the 2006 actual PI score, and did the same for the predicted scores, and then plotted them in Figure 5.3 for the 608 districts with complete data across all seven years to examine the distribution of performance in Ohio. Figure 5.3 provides a means to examine which districts significantly outperform or underperform based on their demographic and context variables. Here, districts separate into three groups, with fifteen districts lying above the 95% confidence interval. We conducted the same procedure with the Texas dataset and identified 32 districts lying above the 95% confidence interval, as shown in Figure 5.4. Each dot represents a district, and districts are labeled by their state AYP accountability rating for the final year of the model. As the model controls for context, demographics, and finance variables, we argue that these districts should be considered for more in-depth analysis, especially qualitative case studies, as our results would suggest that these districts outperform other similar districts and perhaps may be considered effective, equitable and efficient as they outperform other similar districts in the state when controlling for these three types of variables.

**TABLE 5.3** Parameter Estimates for a Hierarchical Linear Growth Model of Ohio and Texas PI Score Change 2004-05 – 2010-11

| | *Ohio PI score change 2005-06 – 2011-12* | | | *Texas PI score change 2004-05 – 2010-11* | | |
|---|---|---|---|---|---|---|
| *Variable* | *Coeff.* | *(SE)* | *Stand. coeff.* | *Coeff.* | *(SE)* | *Stand. coeff.* |
| Intercept | 96.101*** | (0.307) | - | 89.684*** | (0.444) | - |
| Year | 0.511*** | (0.049) | 0.081 | 1.445*** | (0.083) | 0.205 |
| *Intercepts* | | | | | | |
| %Disadvantage students | –1.248*** | (0.109) | –0.199 | –2.182*** | (0.170) | –0.310 |
| %Asian students[a] | 0.689*** | (0.138) | 0.110 | 0.590*** | (0.142) | 0.083 |
| %African American students[a] | –1.401*** | (0.137) | –0.223 | –1.298*** | (0.151) | –0.184 |
| %Hispanic students[a] | 0.266* | (0.170) | 0.042 | –2.034*** | (0.192) | –0.288 |
| % High mobility students | –0.095 | (0.067) | | –0.558*** | (0.097) | –0.079 |
| % Disabled students | –0.601*** | (0.098) | –0.096 | –0.420*** | (0.102) | –0.060 |
| Extra-small enrollment | - | - | - | –1.104*** | (0.390) | –0.077 |
| Small enrollment | –0.181 | (0.262) | | –0.1192 | (0.297) | |
| Large enrollment | 0.521* | (0.261) | 0.038 | –0.134 | (0.386) | |
| Extra-large enrollment | 0.087 | (0.775) | | –0.427 | (0.603) | |
| Urban | –6.499*** | (0.928) | –0.186 | 0.580 | (0.636) | |
| Small Town | –2.178*** | (0.537) | –0.115 | 1.272*** | (0.500) | 0.064 |
| Rural | –1.517*** | (0.398) | –0.120 | 0.951* | (0.457) | 0.066 |
| %Student attendance | 0.740*** | (0.087) | 0.118 | 0.469*** | (0.103) | 0.067 |
| School-level PI score variance | –0.044 | (0.052) | | –0.249*** | (0.161) | –0.036 |
| Avg. teacher years experience | 0.417*** | (0.078) | 0.067 | 0.259*** | (0.114) | 0.036 |
| Student–teacher ratio | –0.227* | (0.089) | –0.036 | 1.250*** | (0.216) | 0.177 |
| Avg. teacher salary | –0.169 | (0.131) | | 2.189*** | (0.121) | 0.310 |
| Instruction expenditure per pupil[a] | 0.175 | (0.142) | | –0.821 | (0.833) | |
| General Administration expenditure per pupil[a] | –0.267*** | (0.088) | –0.043 | –0.693*** | (0.252) | –0.098 |
| School Administration expenditure per pupil[a] | –0.033 | (0.092) | | 0.379*** | (0.173) | 0.054 |
| Instruction Staff Support Service expenditure per pupil[a] | 0.131 | (0.106) | | 0.827 | (0.759) | |
| Capital Outlay expenditure per pupil[a] | –0.111 | (0.062) | | –0.057 | (0.171) | |
| *Slopes* | | | | | | |

| | | | | | | |
|---|---|---|---|---|---|---|
| %Disadvantage | 0.281*** | (0.028) | 0.140 | 0.029 | (0.033) | |
| %Asian | −0.067* | (0.030) | −0.034 | −0.012 | (0.029) | |
| %African American | −0.193*** | (0.029) | −0.097 | 0.146*** | (0.027) | 0.073 |
| %Hispanic | −0.009 | (0.021) | | 0.160*** | (0.034) | 0.080 |
| %Mobile rate | −0.003 | (0.017) | | 0.052* | (0.025) | 0.026 |
| %Disabled students | −0.046* | (0.022) | −0.023 | 0.050 | (0.026) | |
| Extra-small enrollment | - | - | - | −0.040 | (0.085) | |
| Small enrollment | 0.056 | (0.050) | | −0.020 | (0.066) | |
| Large enrollment | −0.037 | (0.055) | | 0.052 | (0.073) | |
| Extra-large enrollment | −0.050 | (0.148) | | 0.213 | (0.114) | |
| Urban | −0.047 | (0.121) | | −0.196 | (0.101) | |
| Small town | 0.015 | (0.070) | | −0.287*** | (0.079) | −0.052 |
| Rural | 0.112* | (0.051) | 0.028 | −0.160* | (0.073) | −0.040 |
| %Student attendance | −0.024 | (0.021) | | 0.025 | (0.024) | |
| School-level PI score variance | −0.053*** | (0.016) | -0.027 | 0.006 | (0.016) | |
| Avg. teacher experience | −0.087*** | (0.018) | -0.044 | −0.019 | (0.025) | |
| Student–teacher ratio | 0.029 | (0.020) | | −0.134** | (0.050) | −0.067 |
| Avg. teacher salary | 0.186*** | (0.031) | 0.093 | −0.377*** | (0.026) | −0.189 |
| Instruction expenditure per pupil[a] | 0.081* | (0.033) | 0.041 | −0.391* | (0.173) | −0.199 |
| General Administration expenditure per pupil[a] | 0.040 | (0.022) | | −0.026 | (0.055) | |
| School Administration expenditure per pupil[a] | −0.005 | (0.020) | | −0.035 | (0.039) | |
| Instruction Staff Support Service expenditure[a] | 0.008 | (0.021) | | 0.326* | (0.159) | 0.0163 |
| Capital Outlay expenditure per pupil[a] | 0.012 | (0.016) | | 0.049 | (0.043) | |
| Amount of Variance between districts (ICC) (%) | 73.16 | | | 50.49 | | |
| Variance Explained | | | | | | |
| Level1: within districts (%) | 16.07 | | | 12.940 | | |
| Level2: between districts (%) | 68.87 | | | 60.039 | | |

*Notes*:

a: % Asian, African American, Hispanic student, Instruction expenditure per pupil, General Administration expenditure per pupil, School Administration expenditure per pupil, Instruction Staff Support Service expenditure per pupil, Capital Outlay expenditure per pupil are square root transformed.

All variables are standardized (z-score) except for the enrollment and locale dichotomous variables.

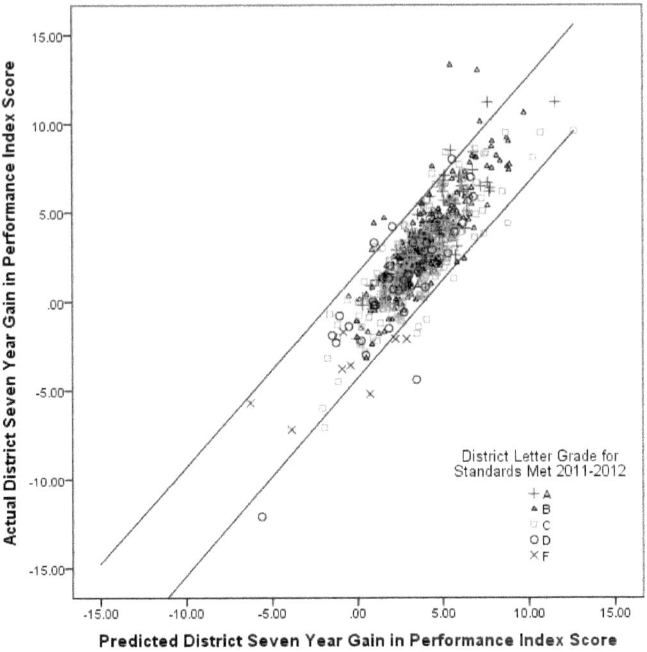


**FIGURE 5.3** Comparison of Predicted versus Actual Seven-year Performance Index (PI) Score Gain in Ohio, 2005-06 – 2011-12.

Hierarchical linear growth modeling predicted PI score gains are plotted on the x-axis for each of the 608 districts in Ohio over the time period, versus the actual gains from 2005-06 to 2011-12. The 95% confidence intervals are shown. Individual districts are plotted as symbols representing the Ohio Department of Education for district accountability rating, with A being the best as Exemplary Performance and F as Academically Unacceptable. Districts that are outside 95% confidence interval are designated as significantly outperforming or underperforming their demographic and context variables.

Finally, we created a quadrant plot comparing the findings of this study to the state accountability ratings for both Ohio and Texas, which are displayed in Figures 5.5 and Figure 5.6 separately. In Figures 5.5 and 5.6, the y-axis indicates the model predicted results, which we defined as PI Gain Index. This was calculated by subtracting the model predicted gain from the actual seven-year PI score gain. The mean of the PI gain index is -0.47 for Ohio and -0.43 for Texas, indicating the average actual seven-year gain is slightly larger than the model predicted seven-year PI gain for both states. The x-axis is the percentage of standards met adopted by Ohio in Figure 5.5 while it is the four-year high school completion rates for Texas in Figure 5.6, which are the main indications of Annual Yearly Progress (AYP) for both states. The vertical dashed line is the average percentage of the standards met for the districts across the state and the two horizontal dashed lines are the 95% confidence interval boundary of the PI

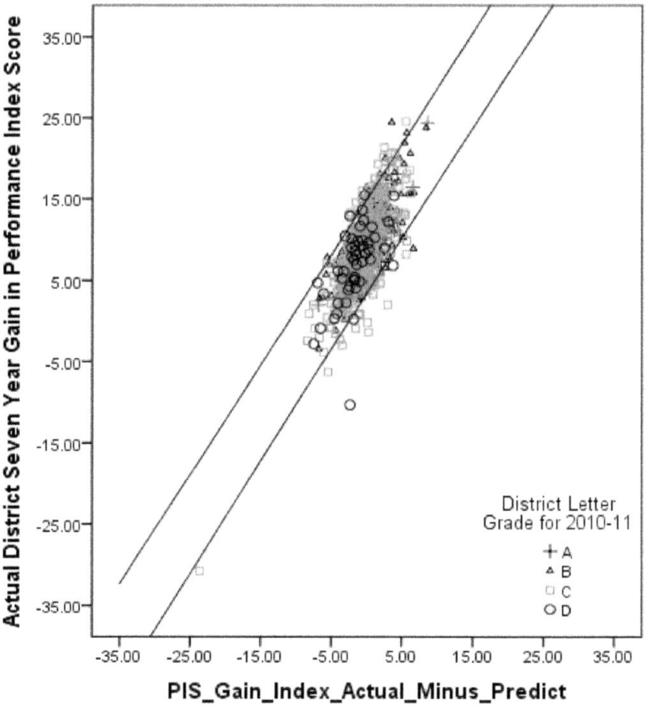


**FIGURE 5.4** Comparison of Predicted versus Actual Seven-year Performance Index (PI) Score Gain in Texas

Hierarchical linear growth modeling predicted PI score gains are plotted on the x-axis for each of the 1,028 districts in Texas over the time period, versus the actual gains from 2004-05 to 2012-13. The 95% confidence intervals are shown. Individual districts are plotted as symbols representing the Texas Department of Education for district accountability rating, with A being the best as Exemplary Performance and D as Academically Unacceptable. Districts that are outside 95% confidence interval are designated as significantly outperforming or underperforming their demographic and context variables.

Gain Index. We also set the markers by the district letter grade for 2010–2011 rated by the Ohio Department of Education (ODE) in Figure 5.5 and district rating grade for 2011–2012 by the Texas Education Agency (TEA). Thus, the three dashed lines separate the districts outside the 95% confidence interval of PI Gain Index into four groups, where we defined the upper left as low-high; upper right as high-high; lower left as low-low and lower right as high-low. Districts in the high-high quadrant (top right) outperform the model here and are rated highly by their state, and thus the two modeling frameworks agree. For the low-low (bottom left) the same can be said as the models agree. However, for the districts in the low-high quadrant (upper left) the model suggests that they should be outperforming, yet by the state accountability indicator they are seen as having less than average performance. The districts in the middle of the plot but above the state averages on the x-axis (far right, middle) are also an interesting group as the model would say that these districts are at the norm when controlling for their contexts and demographics, yet the state rates them very highly. This is mostly likely due to the model controlling for variables such as poverty.

We argue here that the quadrant plots provide a new means for practitioners, researchers and policy makers to identify outperforming or underperforming long-term district performance trends in their states, and surface interesting differences between the state AYP accountability rating system and a more controlled modeling framework such as the hierarchical linear growth model. Note that our argument here is not to replace the state accountability systems; instead, this set of models and visualization is meant to invite further investigation, and help districts bring together central office personnel for professional development, in which district contexts can be matched across HLM growth models (low in one, high in the other, or low in both), as the HLM approach takes advantage of long-term growth while most of the state systems use single year performance for AYP status.

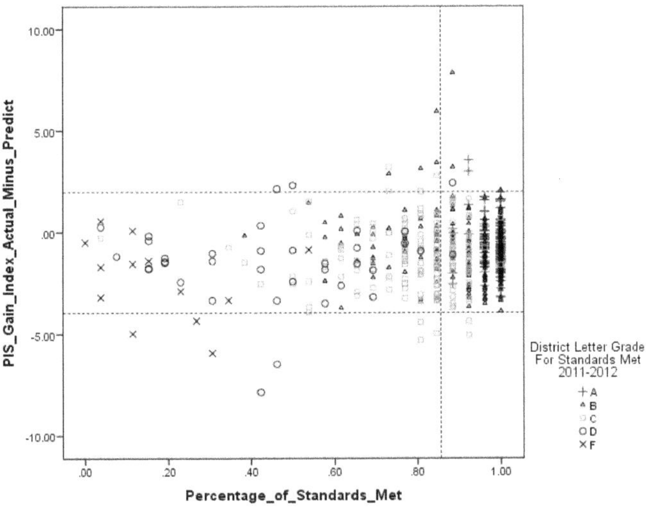


**FIGURE 5.5** Quadrant Plot for Model Results versus Ohio Accountability Ratings.

The difference between hierarchical linear growth modeling predicted PI score gains and actual gain are plotted on the y-axis for each of the 608 districts in Ohio over the time period, versus the percentage of standards met on the x-axis. The 95% confidence intervals are shown as dotted lines. Individual districts are plotted as symbols representing the Ohio Department of Education district accountability rating, with A being the best as Exemplary Performance and F as Academically Unacceptable.

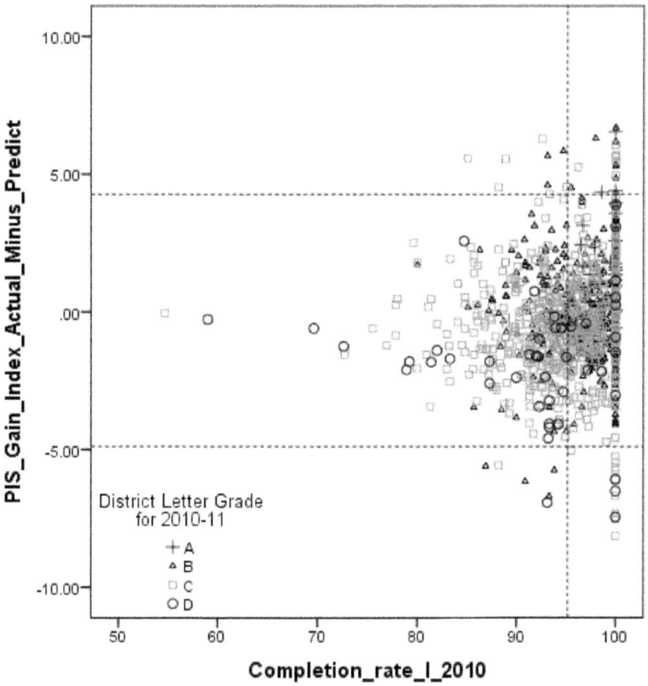


**FIGURE 5.6** Quadrant Plot for Model Results versus Texas Accountability Ratings.

The difference between hierarchical linear growth modeling predicted PI score gains and actual gain are plotted on the y-axis for each of the 1028 districts in Texas over the time period, versus the 4-year high-school completion rates for 2010–2011 on the x-axis. The 95% confidence intervals are shown as dotted lines. Individual districts are plotted as symbols representing the Texas Department of Education for district accountability rating, with A being the best as Exemplary Performance and F as Academically Unacceptable.

## Conclusion

This study demonstrates the utility of using a systems-level analysis of organizations across entire states to understand which school districts in a U.S. state are outperforming similar districts when controlling for multiple variables that are outside the control of the district administration, including context and background variables. Here, we have shown that within the District Effectiveness Research (DER) framework, the hierarchical linear growth modeling strategy controlling for context, demographics and finance variables can work well to identify a set of districts that may be "unusually effective," demonstrating the model in two very different states, Ohio and Texas. As noted in the previous literature in this domain (Bowers, 2008, 2010, 2015), these "outperforming" districts then must be studied using in-depth qualitative studies to understand how the districts outperform other districts in their state that are in similar community contexts. Additionally, comparisons made between state-determined ratings and HLM results have the potential for better understanding of state methods for identifying under- and outperforming districts.

## *Understanding and Improving State Accountability Systems*

The purpose of the analyses presented in this chapter, namely using HLM in this way to identify outperforming districts, is different from that of state-determined ratings. Due to the fact that states not only have the common purpose of identifying schools in need of supplemental support as defined by federal law, they also need to consider a wide range of stakeholders as well as political context. It is for this reason that we do not expect that districts identified as outperforming by the HLM would necessarily be similar to state ratings. Furthermore, because the HLM tracks progress over time, it is possible that districts with high populations of historically underperforming students that are identified as overachieving by the HLM have improved as a result of the additional, targeted support from states. In some cases, it is possible the support was not enough to improve the state's calculation for the district, but enough to grow their population over time. Figures such as 5.5 and 5.6, however, can provide valuable insight to state departments of education as they implement and refine their accountability and school improvement systems under ESSA, such as examining what are the common characteristics across districts within each quadrant of figures such as 5.5 and 5.6. This leads to understanding why a district would be considered outperforming by the HLM, but poorly rated by the state's system, for example. Investigations such as this can help the state calculations identify schools and districts that better reflect initial intentions of the system. Changes to state calculations could include creating conditions that honor growth revealed by the HLM, or excluding indicators unintentionally causing disagreement of the two methods.

## *District Collaboration and Further Qualitative Study*

Despite the abundance of information drawn from the many variables considered in these analyses, this type of modeling falls short of providing reasons for which some districts excel despite consisting of historically underperforming populations of students while others continue to fall farther behind. The final plots presented in Figures 5.3–5.4 and 5.5–5.6 demonstrate a possible next step in this framework for district effectiveness research. For the 15 districts identified as outperforming in Ohio and the 32 districts in Texas, we propose here that the next step in helping districts across a state improve is to gather outperforming districts to begin to identify possible generalizable improvement approaches for further investigation and qualitative deep-dives.

One approach to further investigation is to convene central office staff of the outperforming districts to attend a multi-day work session either through a district or state-led approach. During the convening, attendees could be provided with interactive discussion events in which the discussion is guided to surface initial ideas on how each district finds success in its context, and what is idiosyncratic to that context, and what may be generalizable. This type of conversation can mirror similar types of data and evidence-based conversations and protocols currently being used to help teachers in schools develop improvement strategies around the evidence and processes in their schools (Bowers, 2017). This type of gathering can serve three very important roles. First, central office leaders can build a network of colleagues across the state who are doing similar work and finding similar success. Second, understanding what may be idiosyncratic to the local context is important, as this type of finding can be very helpful to other districts that may not be as successful, but may have experienced similar challenges in a similar or related context. Third, gathering the collective learning from across the top outperforming districts can not only help the other districts across the state know what the similar processes are that are shared across these districts that may lead to increased performance, but the state department of education and the legislature can gain a valuable set of allies in helping in state-level instructional improvement and policy. These districts are most likely to have figured out how to best leverage the policies for helping their students, as previous research has noted (Bowers, 2008; Skrla et al., 2011; Thompson et al., 2008).

For further study, districts at or below the norm could pair with outperforming districts with similar student populations to understand processes and foci in order to learn what is transferrable. State departments may not always have resources to facilitate this, and therefore district collaboration does not necessarily need to be led at the state level. For example, districts may develop programs to share their successes with other districts in the state in which other districts may send staff to visit selected classrooms and hold in-depth conversations with school leadership. Overall though, after such workshops (either at individual districts or larger gatherings across districts with state administration) and reporting the results to the state, we encourage states to consider pairing districts at the norm with one of the outperforming districts, to have the central office staffs and principals meet periodically over multiple years, tour each other's campuses, and discuss how their operations and instructional focus may differ, and how they may be able to collaborate. There may be much to learn from each of the district locations, and district central office administrators rarely have an opportunity to engage over a long period of time in discussing these types of differences and possible improvement activities and professional development. In this way, districts across the system may inform their learning and practices through these types of systematic evidence-informed collaborative opportunities.

## *Limitations*

While we argue that our results are robust, this study is limited in multiple ways. First, while we included two states with different contexts and seven years of data for each state, this study thus does not consider the extent to which this modeling framework may perform in the other 48 states in the U.S., or across nations. Second, while the longitudinal data included seven years of data, the final years of the models were 2011–2012 (Ohio) and 2010–2011 (Texas). These end dates were selected as for each state the testing regime changed, as the states transitioned to new assessment procedures and new standardized tests across subjects, making it difficult to study longitudinal performance beyond these dates. This necessarily focuses the models on the first decade of the twenty-first century. We encourage future research to consider more recent data that includes all of the years since, as well as to consider how to adapt the models to perhaps account for the test changes over time. Third, our definition of effectiveness of calculating the Performance Index Score, while an improvement over previous district effectiveness research that focused on just cross-sectional English or mathematics test scores, still consists of state standardized assessments, even if these include many grade levels and subjects over time. We encourage future research to examine district effectiveness through broader conceptions of "effective" beyond test scores, including student socio-emotional learning and grades, student graduation from high school, college admission, college degree completion, citizenship, community engagement and volunteerism, and career training, readiness, and placement. Examining the difference in which districts outperform on these broader measures of effectiveness across a state could provide valuable insights into how to successfully serve students and communities across a broad range of potential positive schooling outcomes.

## References

Bowers, A. J. (2008). Promoting excellence. *Leadership and Policy in Schools*, 7(2), 154–177.

Bowers, A. J. (2010). Toward addressing the issues of site selection in district effectiveness research: A two-level hierarchical linear growth model. *Educational Administration Quarterly*, *46*(3), 395–425.

Bowers, A. J. (2015). Site selection in school district research: A measure of effectiveness using hierarchical longitudinal growth models of performance. *School Leadership & Management*, *35*(1), 39–61.

Bowers, A. J. (2017). Quantitative research methods training in education leadership and administration preparation programs as disciplined inquiry for building school improvement capacity. *Journal of Research on Leadership Education*, *12*(1), 72–96.

Bowers, A. J., & Lee, J. (2013). Carried or defeated? Examining the factors associated with passing school district bond elections in Texas, 1997–2009. *Educational Administration Quarterly*, *49*(5), 732–767.

Bowers, A. J., & Urick, A. (2011). Does high school facility quality affect student achievement?: A two-level hierarchical linear model. *Journal of Education Finance, 37*(1), 72–94.

Bowers, A. J., & White, B. R. (2014). Do principal preparation and teacher qualifications influence different types of school growth trajectories in Illinois? A growth mixture model analysis. *Journal of Educational Administration, 52*(5), 705–736.

Bryk, A. S. (2010). Organizing schools for improvement. *Phi Delta Kappan, 91*(7), 23.

Card, D., & Payne, A. A. (2002). School finance reform, the distribution of school spending, and the distribution of student test scores. *Journal of Public Economics, 83*(1), 49–82.

Chingos, M. M., Whitehurst, G. J., & Gallaher, M. R. (2015). School districts and student achievement. *Education Finance and Policy, 10*(3), 378–398. https://doi.org/10.1162/EDFP_a_00167

Condron, D. J., & Roscigno, V. J. (2003). Disparities within: Unequal spending and achievement in an urban school district. *Sociology of Education, 76*(1), 18–36.

Daly, A. J., & Finnigan, K. S. (2012). Exploring the space between: Social networks, trust, and urban school district leaders. *Journal of School Leadership, 22*(3), 493–530.

Daly, A. J., & Finnigan, K. S. (Eds.). (2016). *Thinking and acting systemically: Improving school districts under pressure* (Reprint edition). Washington, DC: Amer Educational Research Assn.

Elmore, R. F., & Burney, D. (1997). *Investing in teacher learning: Staff development and instructional improvement in community school district # 2, New York City*. Retrieved from http://eric.ed.gov/?id=ED416203

Elmore, R. F., & Burney, D. (1999). Investing in teacher learning: Staff development and instructional improvement. In L. Darling-Hammond & G. Sykes (Eds.), *Teaching as the learning profession: Handbook of policy and practice* (pp. 263–291). San Francisco: Jossey-Bass.

Ferguson, R. F. (1991). Racial patterns in how school and teacher quality affect achievement and earnings. *Challenge, 2*(1), 1–35.

Finn, J. D., Gerber, S. B., & Boyd-Zaharias, J. (2005). Small classes in the early grades, academic achievement, and graduating from high school. *Journal of Educational Psychology, 97*(2), 214–223.

Firestone, W. A., Mangin, M. M., Martinez, M. C., & Polovsky, T. (2005). Leading coherent professional development: A comparison of three districts. *Educational Administration Quarterly, 41*(3), 413–448.

Hallinger, P., Heck, R. H., & Murphy, J. (2014). Teacher evaluation and school improvement: An analysis of the evidence. *Educational Assessment, Evaluation and Accountability, 26*(1), 5–28

Hanushek, E. A. (1989). The impact of differential expenditures on school performance. *Educational Researcher, 18*(4), 45–62.

Heck, R. H., Thomas, S. L., & Tabata, L. N. (2013). *Multilevel and Longitudinal Modeling with IBM SPSS* (2nd ed.). New York, NY: Routledge.

Honig, M. I. (2003). Building policy from practice: District Central office administrators' roles and capacity for implementing collaborative education policy. *Educational Administration Quarterly, 39*(3), 292–338.

Honig, M. I. (2008). District central offices as learning organizations: How sociocultural and organizational learning theories elaborate district central office administrators' participation in teaching and learning improvement efforts. *American Journal of Education, 114*(4), 627–664.

Honig, M. I., & Rainey, L. R. (2015). *How school districts can support deeper learning: The need for performance alignment. Deeper Learning Research Series*. Retrieved from http://eric.ed.gov/?id=ED560756
Hox, J. (2010). *Multilevel analysis: Techniques and applications* (2nd ed.). New York: Routledge.
Leithwood, K. (2010). Characteristics of school districts that are exceptionally effective in closing the achievement gap. *Leadership and Policy in Schools, 9*(3), 245–291.
Levin, B., Datnow, A., & Carrier, N. (2012). Changing school district practices. Retrieved from https://studentsatthecenterhub.org/wp-content/uploads/2010/01/Changing-School-District-Practices.pdf
Murphy, J., Hallinger, P., & Heck, R. H. (2013). Leading via teacher evaluation: The case of the missing clothes? *Educational Researcher, 42*(6), 349–354.
NCES. (n.d.). Common Core of Data. Retrieved from http://nces.ed.gov/ccd/
Ni, X., Bowers, A. J., Esswein, J. (2016). *What counts in calculating school and district level performance index scores: A summary and analysis of academic performance index metrics across the 50 states*. New York: Teachers College.
O'Connor, J., & McDermott, I. (1997). *The art of systems thinking: Essential skills for creativity and problem solving*. San Francisco, CA: Thorsons.
ODE (2014). Office of data services. Columbus, OH: Ohio Department of Education Retrieved from http://education.ohio.gov/
Okpala, C. O., Okpala, A. O., & Smith, F. E. (2001). Parental involvement, instructional expenditures, family socioeconomic attributes, and student achievement. *The Journal of Educational Research, 95*(2), 110–115.
Opfer, V. D., Henry, G. T., & Mashburn, A. J. (2008). The district effect: Systemic responses to high stakes accountability policies in six southern states. *American Journal of Education, 114*(2), 299–332.
Payne, K. J., & Biddle, B. J. (1999). Poor school funding, child poverty, and mathematics achievement. *Educational Researcher, 28*(6), 4–13.
Purkey, S. C., & Smith, M. S. (1985). School reform: The district policy implications of the effective schools literature. *The Elementary School Journal, 85*(3), 353–389.
Senge, P. (1990). *The fifth discipline: The art and practice of the learning organization*. New York: Currency, Doubleday.
Skrla, L., McKenzie, K. B., Scheurich, J. J., & Dickerson, K. (2011). Home-town values and high accountability: A Texas recipe for districtwide success in an urban school district. *Journal of Education for Students Placed at Risk (JESPAR), 16*(2), 137–165.
Smylie, M. A. (2009). *Continuous school improvement*. Thousand Oaks, CA: Corwin.
Thompson, C. L., Sykes, G., & Skrla, L. (2008). *Instructionally-focused district leadership: Toward a theoretical account*. Retrieved from http://education.msu.edu/epc/forms/Thompson_et_al_2008_Coherent_Instructionally_Focused_District_Leadership.pdf
Trujillo, T. (2013). The reincarnation of the effective schools research: rethinking the literature on district effectiveness. *Journal of Educational Administration, 51*(4), 426–452.
Trujillo, T. (2016). Learning from the past to chart new directions in the study of school district effectiveness. In A. J. Daly & K. S. Finnigan (Eds.). *Thinking and acting systemically: Improving School districts under pressure* (pp.11–48). Washington, DC: Amer Educational Research Assn.
Verstegen, D. A., & King, R. A. (1998). The relationship between school spending and student achievement: A review and analysis of 35 years of production function research. *Journal of Education Finance, 24*(2), 243–262.

# 6

# WHOLE DISTRICT TRANSFORMATION

## Leading Systems Change for Sustainability

*Cynthia L. Uline and Lisa A.W. Kensler*

The largest social movement in history, comprising individuals and organizations in for-profit and non-profit sectors worldwide, is currently advancing the work of sustainability, aiming to insure the highest quality of life for the global community (Hawken, 2007). Generally speaking, sustainability captures the interface between human civilization and the natural world and emphasizes the degree to which human beings choose to live within the ecological carrying capacity of planet earth, presently and into the future. Sustainability science applies a systems thinking approach to examine planetary, social, and economic issues in a holistic fashion, exploring the question of how humans might live on planet earth in ways that are more ecologically healthy, socially just, and economically viable (Cassell & Nelson, 2010; Goleman & Senge, 2014; Williams, Kennedy, Philipp, & Whiteman, 2017).

Formal PK-12 education plays a critical role in the evolution of our social systems by introducing content knowledge and modeling social habits and practices. In terms of the sustainability movement, greener schools advance a relevant and integrated curriculum (Henderson & Tilbury, 2004; Nolet, 2015, 2009) and model sustainable building design and management practices for students and the broader community (Edwards, 2006; Higgs & McMillan, 2006). Educational leaders throughout the United States and the world are engaged in sustainability-related efforts through a variety of green school practices, including recycling, reduced energy consumption, toxic-free cleaning methods, intentional purchasing, green building design features, and the like (Gordon, 2010; Henderson & Tilbury, 2004; Huckle & Wals, 2015; Kensler & Uline, 2017; Stone, 2009; Wu, 2002). And, as these sustainability-focused school leaders move in the direction of whole school and whole systems approaches to sustainability, they not only teach about sustainability via the curriculum, but the school community becomes a vibrant place for learning together how to live more sustainably (Birney & Reed, 2009; Rauch, 2002; Stone & Barlow, 2005). Processes such as designing, building, and managing healthy, high performance facilities (Bernstein, 2003; Gordon, 2010), purchasing, reusing, and recycling ecologically intelligent materials (Goleman, 2009), and growing and serving more local, organic food (Morgan & Sonnino, 2008) become engaging learning resources and opportunities for adults and children.

The research described here sought in-depth understanding of sustainability-focused, system-wide reform efforts within a large urban school district in the United States. This school district received three years of support from a leading national, sustainability-focused organization, the Center for Sustainability Leadership's (CSL) (pseudonym), Sustainability Champion (SC) Program (pseudonym). The SC program provided school districts with a partially-funded, dedicated staff person (the SC) who worked alongside district leaders, serving as the sustainability contact person for three years. The SC worked hands-on with principals, teachers, staff, and students, facilitating the implementation of sustainable practices district-wide. Grant funding allowed CSL to employ the SC for a two-year period and share the employment costs with the school district during year three. Beginning with district-defined metrics for success, the SC, operating on a daily basis within the school district, facilitated district-wide sustainability or "greening" efforts through provision of expert information, training, and guidance. To guide efforts, the SC herself received ongoing, systems change-related professional development by the CSL. Our research followed and documented system-wide change strategies and related outcomes over three years.

Our literature review introduces two fundamental principles for leading schools as the living systems they are: diverse networks of relationships and self-organizing, self-regenerating systems. Following this discussion of leadership principles will be a more detailed explanatory review of what it means to be a school practicing whole school sustainability. The study's methodology, findings, and discussion follow.

## Literature Review

The aim of the change initiative studied here was to move a school district along a continuum towards whole school sustainability. As described in Kensler and Uline (2017), whole school sustainability attends to both the social and ecological systems of a school. Democratic and ecological principles inform the cultivation of healthy social and ecological circumstances within a school's built and natural environment. When leaders understand their schools as living systems, their strategies shift to better align with this whole system thinking.

### *Living Systems Understandings Inform Organizational Change for Sustainability*

Sustainability calls for the intentional and intelligent integration of human and ecological systems. It is both a theoretical and practical field with a dramatic sense of urgency and relevance to all communities. Systems thinking drives both the science and the practice, as scholars and practitioners across a wide array of disciplines, including the life and physical sciences, engineering, economics, and social sciences (Bettencourt & Kaur, 2011), explore the ways in which "sustainability is . . . a property of an entire web of relationships . . . involv[ing] a whole community" and extending out into the attendant communities (Capra, 2009, p. 2). As Capra observed, "A sustainable human community interacts with other communities – human and nonhuman – in ways that enable them to live and develop according to their nature" (Capra, 2009, p. 2).

**Diverse systems of networks and relationships.** The sustainability movement is very much about better aligning human ways of living on this planet with the laws of nature so that humans and nature might thrive into the future. For the sake of scientific accuracy and conceptual coherence, it makes sense that educators, interested in leading sustainability-focused reforms, extend their systems thinking habits of mind to consider what healthy and sustainable life systems, the systems in which our social systems exist and upon which they depend, can teach us (Capra, 1996, 2002; Wheatley, 1999; Wheatley & Kellner-Rogers, 1996; Wielkiewicz & Stelzner, 2005). The most resilient and successful living systems are diverse communities with a plethora of partnerships set within dense networks of relationships (Capra, 1996, 2002). Ecologists have documented that communities with greater species diversity, partnerships, and networks are better able to withstand and recover from major disturbances (Peterson, Allen, & Holling, 1998).

This ecological principle of networks, partnership, and diversity relates directly to our social systems as well. Diverse partnerships within and outside school district organizations strengthen school districts' and schools' capacity to innovate and become a central hub of resources, education, and learning (Blank, Jacobson, & Melaville, 2012; Daly & Finnigan, 2010; Heers, Van Klaveren, Groot, & Maassen van den Brink, 2016; Reina et al., 2014; Satullo, 2012).

**Self-organizing, self-regenerating systems.** Living systems have been self-organizing since the beginning of life itself. Although we may not fully understand how life began on planet earth, scientists have learned a lot about how living systems organize, grow, and change. Self-organization leads to the emergence of novelty and adaptation (Capra, 1996, 2002). Capra (2010) explained:

> Living systems generally remain in a stable state, even though energy and matter flow through them and their structures are continually changing. But every now and then such an open system will encounter a point of instability, where there is either a breakdown or, more frequently, a spontaneous emergence of new forms of order . . . [This emergence] has been recognized as the dynamic origin of development, learning, and evolution. In other words, creativity – the generation of new forms – is a key property of all living systems.
>
> *(n.p.)*

Living systems are open and closed systems. As Capra explained, they are open to their environment, using energy and matter to maintain their structure. However, they are operationally closed, meaning they contain the information that they need to be and reproduce themselves. He described this fundamental characteristic of life as "self-maintenance via a mechanism of self-regeneration from within" (Capra & Luigi, 2014, p. 131). As environmental conditions change, life responds creatively. Understanding this principle can help inform how we think about change within our organizations. Rather than imposing change from positional power atop the organizational chart, leaders, who apply a living systems perspective to their work, cultivate the conditions within which change happens and novelty emerges. These conditions thrive on diverse networks characterized by high levels of trust and power sharing.

Those in positional power have the choice to share their power or not. They can view power as a limited commodity, hoard it, and strive to keep control, leading to less flexible, less trusting hierarchies and bureaucracies. Social systems like these are unhealthy, less productive, and not characterized by continuous learning or improvement; their social networks are more rigid and less resilient (Daly, 2009; Daly, 2010; Daly, Liou, Tran, Cornelissen, & Park, 2013). When positional leaders choose to share or distribute their power by empowering others, active networks of relationships purposefully form and dissipate as needed, in order to accomplish necessary work (Wheatley, 1999).

> This strategy capitalizes on the fact that key insights often emerge in unusual places. By working through organized networks, the likelihood increases that these ideas may surface, be systematically examined, and if promising, moved rapidly into testing and refinement.
>
> *(Bryk, Gomez, Grunow, & LeMahieu, 2015, p. 11)*

Recent studies into the formation, dissolution, and evolution of such flexible and supportive capacity-building networks between teachers, principals, district administrators, using a social capital or social network perspective, have deepened our understanding of the role social interactions, interdependency, and interconnectedness play within and across schools as educators seek to implement reforms, advance professional learning, and improve student learning (Coburn & Russell, 2008; Daly, 2010; Scott, 2013)

### *Whole School Sustainability Modeled According to Networked, Self-organizing Systems*

Green schools are education's contribution to the sustainability movement. Across the United States, 85,530 traditional public, 4,480 public charter, and 26,230 private schools (NCES, 2013) present a significant opportunity for PK-12 education to assist in creating more sustainable socio-ecological systems, locally and globally. As educators work to green existing school facilities, and apply green principles to the design and construction of new schools, they articulate and advance sustainability goals and purposes. Green school facilities dramatically reduce negative environmental impacts and improve occupants' well-being, while engaging children and adults in learning about these efforts. Students and teachers can experience the school building as a three-dimensional textbook, modeling socio-ecologically aware norms and practices (Kong, Rao, Abdul-Rahman, & Wang, 2014; Taylor, 2009). Beyond conserving energy, decreasing stress on natural resources, preserving surrounding habitats, reducing waste, and attending to well-being, the ecological literacy of our students, teachers, administrators, and community members improves.

Green schools and the practice of whole school sustainability are relatively new, and rapidly growing, phenomena in PK-12 education led by a new generation of school leaders who are applying sustainability-focused frames of mind and practice to all aspects of their work (See for example, Kensler & Uline, 2015; Kensler & Uline, 2017; Scott, 2013). Leading schools and districts for sustainability (green schools) requires intentionally and systematically re-designing our educational systems and processes from the inside out. Adding a couple of green activities here and a few sustainable practices there, only creates a distracting layer of activity and does little to change the essential character of education. Think of these as veneers, more coats of paint, whatever the color.

> In schools aspiring to be a darker shade of green . . . woven into the school ethos . . . [is] the development of whole-school policies and practices to foster environmental concern and responsiveness, but also to promote social justice, multiculturalism, holistic health, safe schools, citizenship, and democracy across the school community.
>
> *(Selby, 2000, p. 91)*

The very goals of sustainability can "redefine the role of schools and their relationship with the community . . . [rendering] schools as a focal point where children, adults, and the community interact and learn together" (Henderson & Tilbury, 2004, n.p.). According to recent counts, communities across the globe benefit from 2,020 certified and 2,134 registered LEED (Leadership in Energy and Environmental Design) PK-12 school projects (US Green Building Council, 2017). Sustainability-related reform efforts are also reflected in the 735 Eco-Schools across 45 U.S. states, plus the District of Columbia (NWF, 2016) and a growing cadre of Green School Alliance members (https://www.greenschoolsalliance.org/) and 340 U.S. Department of Education Green Ribbon School awardees, since 2012 (https://www2.ed.gov/programs/green-ribbon-schools/awards.html).

Although the actual number of green schools is still a very small percentage of schools overall, early exemplars apply whole school sustainability (WSS) as a high leverage strategy for addressing twenty-first-century challenges, from student engagement and performance to climate change and community resilience (Kensler & Uline, 2017). Children have the opportunity to interact in, and with, ecologically responsive learning environments that reveal connections to the larger society and natural world. Through active, problem- and project-based learning experiences, students begin to build a sense of responsibility for, and commitment to, their community, their planet, and each other.

## Study Design and Methods

The study utilized instrumental case study methods (Stake, 1995) to chronicle one large urban school district's responses to an intervention aimed at "greening" the district from the inside out. The district partnered with the Center for Sustainability Leadership (CSL) and engaged the Sustainability Champion (SC), funded by CSL for two years, to work within the district as an internal change agent with significant external support from the CSL. The instrumental case study utilizes the chosen case (i.e. the school district within which the SC was assigned to be an internal change agent) as an opportunity to gain understanding of a particular phenomenon (i.e. the SC's facilitation of school district adoption of ecologically sound practices) (Stake, 1995). The Green and Sustainable School District (GSSD) (pseudonym) was purposefully chosen as a site for this research, because it was an early partner with the CSL in the SC program. We explored key aspects of the partnership and change initiative including the selection process for both the district and SC, the CSL's professional development support for the SC, and the longer-term outcomes related to greening from the inside out.

The following research questions framed the study: 1) What systems change strategies did the school district employ to facilitate systems change within the GSSD?; 2) To what extent were these change strategies supported by the SC and the CSL-provided professional development?; 3) In what ways and to what extent were district-wide, sustainability-focused reforms realized over the three-year period?

## Data Sources

Data were collected via interviews with 40 participants, including the SC, district administrators and staff, principals, teachers, and city government and community partners throughout the three-year period. Researchers employed the Critical Incident Technique (CIT) to capture detailed behavioral descriptions in the context of key real-world situations and occurrences (Query, Kreps, Arneson, & Caso, 2001). The CIT qualitative approach employed explicit interview methods to collect observed incidents having special significance and meeting systematically defined criteria. This interview technique probed where incident(s) happened, the reactions of participants involved in the incident(s), and the result of the described incident(s) (Kvale, 1996). CIT provides opportunity for researchers to "[see] beyond what appear to be isolated and independent incidents to deeper patterns . . . recognizing connections between events" (O'Conner & McDermott, 1997, p. 7). Thus, CIT provides a kind of systems thinking tool in itself, allowing the researcher to "[see] patterns of change rather than static snapshots" (Senge, 1990, p. 68).

The initial timeline of critical incidents began with the events identified by the initial group of key participants. Additional participants were selected based on specific selection criteria, including (a) individuals who were employed by, and/or involved with the school district during the time the district greening efforts occurred, and (b) individuals who were knowledgeable about the associated district goals and strategies. Each subsequent interview participant was made aware of the timeline and encouraged to share additional information about events or activities. They were also asked to add additional items they felt were critical to school district's progress. Through identification and discussion of critical incidents, the research participants provided feedback regarding the effectiveness of SCs and district leadership, the challenges leaders and school district members faced in instituting environmentally sound practices, and the various strategies employed to overcome reform-related obstacles encountered at the personal, school, or district level.

We triangulated data from individual and focus group interviews, field observations, and document analyses, conducted at the case study school district (Denzin, 1970; Merriam, 1998; Patton, 1990; Stake, 1995). The school district facilities department members and the district level Green Team participated in focus group interviews, with all others conducted as individual interviews. Each interview took place in a quiet location (office or conference room) and lasted between 40 and 60 minutes. Documents included meeting agendas, training literature, SC quarterly reports, and electronic communications. In addition, monthly conference calls between the SCs and CSL staff, for purposes of providing ongoing support, were recorded and transcribed for analysis.

We visited the district five times, beginning in the spring of the first school year and following each fall and spring until the end of year-three study. At least 30 hours of interview data were collected, transcribed, and analyzed for this project. Researchers also attended and observed relevant district- and school-level meetings, observed various school-level green activities, and observed six, two and one-half day professional development sessions, conducted outside of the district, that the CSL designed primarily for the SCs' learning needs.

### *Data Management and Analysis*

With respect to qualitative data management and analysis, all interviews were digitally recorded and transcribed. Data analysis occurred continuously throughout data collection as the researchers identified emerging themes, as well as teased out anomalies and contradictions (Merriam, 1998). Preliminary categories were generated from the literature on green schools and altered as additional themes and patterns emerged (Fook, 2002). The use of member checks, peer debriefing, triangulation, as well as the maintenance of an organized documentation system, helped confirm research findings. Corresponding universities' Institutional Review Boards (IRB) reviewed and approved all aspects of the study design and procedures, as did the local school district.

## Findings

Twenty-nine school districts across the United States applied to participate in the Center for Sustainability Leadership's (CSL) Sustainability Champion (SC) Program. The GSSD was one of two districts chosen to participate because of their stated and demonstrated commitment to greening their district across policies, facilities, curriculum, and culture. Following the district selection, an SC for the district was selected from a pool of 66 applicants. The school district's application letter provided evidence of the district's commitment to sustainability, noting the prior adoption of sustainability-related programs. A sampling of these programs includes: model school policy for reducing greenhouse gases; a high school facility modernization project as the district's first high-performance,

**TABLE 6.1** School District Demographic Data

| | |
|---|---|
| Number of schools | 81 |
| Number of students | 47,890 |
| Eligible for free/reduced meals | 69% |
| Race/ethnicity | 36% Hispanic; 18% Asian; 16% Black; 19% White; 7% two or more races/ethnicities |

inner-city construction project; the district's hiring of three full-time energy management supervisors; and a healthy food leadership team who was charged with overseeing a whole-scale transformation of the district food service program including salad bars and school gardens at every school. Table 6.1 reports school district demographic data.

## *Critical Incidents Shaping GSSD Sustainability-focused Reform Efforts*

During the three-year period of the Sustainability Champion (SC) Program, GSSD experienced a series of critical incidents, which participants deemed important to shaping their school district's sustainability-focused reform efforts. Based on analysis of focus group and individual interviews with key informants from the district, we selected eight events as central to the school's reform. These events are briefly described in Table 6.2. The critical incidents are presented in chronological order, followed by a detailed discussion of each.

## *Mayor and Superintendent Sowed Seeds of Reform*

A year prior to applying for the SC program, the city's mayor launched the Green Enterprise (pseudonym) to convene and align leaders in the region and establish a shared vision for sustainability. According to the Mayor's Office Green Enterprise Director, "The Mayor convened more than 275 experts and community leaders during an eight-month initiative and identified challenges and opportunities through five policy committees: Energy, Green & Clean Technology, Urban Design & Green Building, Waste & Recycling and Water & Nature." The school district's CSL application letter described the city's current efforts to build a green economy, calling the city "ground zero for the grassroots green revolution, a place where reduce, reuse, recycle isn't a bumper sticker but a mantra" (n.p).

**TABLE 6.2** Critical Incidents

| *Event Number* | *Description* |
|---|---|
| Critical Incident 1 | Mayor and Superintendent Sowed Seeds of Reform |
| Critical Incident 2 | Engaging Green Launched on Earth Day |
| Critical Incident 3 | District Chosen to Receive SC/SC Chosen and Hired |
| Critical Incident 4 | SC Attended CSL-Provided Professional Development Sessions |
| Critical Incident 5 | SC Reached Out to Potential Partners Within and Beyond the School District |
| Critical Incident 6 | District Established a District-level Green Team |
| Critical Incident 7 | District Drafted a Sustainability-focused Facility Master Plan and Passed Bond Issues Totaling $440 Million |
| Critical Incident 8 | Engaging Green Facilities Projects Implemented |

The GSSD superintendent began his tenure two years prior to submitting the SC program application. Early in his tenure he visited schools and classrooms across the district. He listened, observed, and asked questions.

> In those first 100 days, during lunch at an elementary school, a second grader asked *me* a question, 'Why do we still use Styrofoam trays?' He is now in the sixth grade. I was in his school recently, and I took a picture of him holding one of the compostable trays we now use. We changed the trays because of him. I arrived with no intention of greening schools, but I think you come into a community and, if you are smart, you will listen and learn about the culture. Then [a year later], I got invited to [a national summit on green schools], along with the Mayor. That was the real [eye opener].

From this point forward, the GSSD superintendent served as the primary champion for the school district's sustainability-related initiatives. He was interested in maintaining visibility for all things green and building awareness, interest, and capacity across the district. Green initiatives were also included within the district's strategic plan (e.g. a garden at all school sites). As one district administrator reported, "The superintendent's evaluation will now include these concerns . . . it becomes a more holistic evaluation."

## *Engaging Green Launched on Earth Day*

The SC's work would center on operationalizing the Engaging Green initiative, hands-on, green-focused learning experiences aimed at teaching students to think critically, solve real-life problems, and develop as eco-leaders. The GSSD superintendent described the genesis of Engaging Green.

> At [the national summit] we were able to think with others. That is where we came up with this idea of Engaging Green. Through Engaging Green, we could engage our students in the solution [the adoption of ecologically sound habits and practices] along with their parents and communities, placing schools at the center and the focal point.

Around the same time, the district changed their logo from a red to a green apple, underscoring the district's commitment to sustainability. According to the superintendent, "This symbolic gesture creates a constant daily reminder we are here to serve kids, but also to be good stewards of our resources and of our people and our children."

Under the SC's leadership, campuses across the district were encouraged to form student-led Green Teams to conduct "green" audits of school facilities. With the help of teachers, parents, district facilities staff, and community experts, the teams drafted recommendations for green improvements, ranging from replacing outdated windows to installing low-flow plumbing fixtures. One high school principal spoke about the need to invite district operations and maintenance departments to engage with students in their assessments of building conditions.

> The operations and maintenance folks were so excited to be a part of the process. Some of these gentlemen have been here for forty years. They're craftsmen. We have nine different shops led by nine different foremen who possess specialty trades. This represents an enormous store of knowledge to bring to the table, and nobody had ever asked.

These place- and problem-based projects engaged student voice, as teams analyzed how much energy they consumed, how much water they used, the kinds of materials their buildings were made of, how they handled wastes, what kinds of cleaning and chemicals they used, and presented their recommendations to a panel of local experts from the fields of architecture, engineering, energy, land-use planning, and water management. The panel judged the exhibits, presentations, and written papers using a rigorous scoring rubric. Fifteen schools were awarded funds for these green facilities projects.

The team who scored highest for its presentation and report advanced strong rationale for providing the school with automatic hand dryers, rain barrels, solar tubes and upgrades to irrigation and HVAC systems. The judging panel voted to allocate up to $550,000 for this one school's projects. In all, the first 15 award-winning presentations were allocated $5 million in capital improvement funds (made available through existing bond monies).

### *District Chosen to Receive Sustainability Champion (SC) and SC Chosen and Hired*

GSSD's sought to place Engaging Green at the center of the SC responsibilities. Related short-term goals included the following: establishment of school-level student Green Teams, the training of Green Teams to perform green audits at school sites, development of proposals for improving green audit results to be presented at a science fair-type event, and awarding of funds to implement relevant portions of Engaging Green facilities projects. The SC would also support corresponding communications initiatives to "sell the need," laying the foundation for another grassroots facilities bond to fund the retrofitting, renovation, and greening of schools across the district. Long-term goals included passage of the local bond initiative, work with the state legislature to promote state provision of matching dollars for green projects, and identification of new stakeholders, partners, and revenue streams. With a Bachelor of Science degree in Construction Management and a Master of Science degree in Construction

Management (sustainable building emphasis), as well as six years' experience in the field, the SC came to the job with the skills necessary to ensure students were prepared to conduct energy audits; assist principals and Green Team Advisors in conceptualizing appropriate facility projects; facilitate the judging of the project competition; oversee the bid process for various projects; and assist in managing the construction phase that would bring projects to fruition.

SC shared perceptions of the fit within the district this way:

> I've worked in construction for six years; my background is construction management engineering. [In my interview for this position] I said, "My people are the guys at the end of the hammer."

When the SC arrived at GSSD, she was housed in the facilities department, with the Director of GSSD's Maintenance and Operations division serving as her direct supervisor. At the same time, the district Chief of Staff served as a central point of contact for the SC work, helping her negotiate her new and complex work context.

### *SC Attended CSL-Provided Professional Development Sessions: Learning to Think with a Systems Point of View*

Asking district administrators, maintenance and operations staff, principals, and teachers to shift their fundamental beliefs and practices presents a formidable challenge. For some select departments and schools within the district, implementation of sustainable practices involved refinement and extension of work already under way. For most, these new ways of thinking necessitated a transformation in the norms of practice.

Over the three-year period, the SC traveled to CSL headquarters for semi-annual, in-person training with other SCs from other districts. Winter trainings were specifically tailored to the SCs needs, providing just-in-time learning around concerns specific to their work, including change management, systems thinking, team building, marketing, and problem solving. Summer trainings occurred within the context of annual Sustainability Leaders Institutes (pseudonym), placing the SCs in conversation with district-level sustainability leaders from across the country. Throughout their training, the SCs heard a consistent, learner-centered message, prompting them to advocate for green schools as a means to articulate and advance larger educational purposes.

During the winter of their first year, the SCs were introduced to systems thinking by Linda Booth Sweeney, author of *Systems Thinking Playbook* (1995) and *Connected Wisdom: Living Stories about Living Systems* (2008). The SCs considered the myriad moving and interacting parts of their respective school systems, the inherent organizational complexities, and the multiple, sometimes contradictory, goals across stakeholders. They learned that effective leaders make systems visible, help stakeholders see connections, use language that better mirrors the interdependent and dynamic reality, and identify key leverage points for change.

A handout from a second-year training on organizational change, presented by a sustainability expert from a renowned east coast university, described the session as follows: "We will examine key leverage points for managing organizational change for sustainability: facilitating reform through changes in district contracts, procedures, policies; facilitating behavioral change and utilizing partnerships; and building consensus as a force for change".

During year two of the program, the SC, with CSL support, wrote and released the district's first sustainability report. The report told a compelling story of change, combining a specific account of cost savings to date with descriptions of specific district-wide accomplishments and detailed summary of specific initiatives implemented across 73 of the 81 district schools, including learning gardens, student recycling programs, alternative transportation programs, air quality programs, tree plantings, and wellness councils.

### *SC Reached Out to Potential Partners Within and Beyond the School District*

The SC's mentors within the district, in consort with her supervisor at the CSL, established an early work plan that laid the foundation for establishing the community-wide network necessary for accomplishing the work. The SC reached out to the schools and to the community beyond. Representatives from key community agencies acknowledged their roles within the network, and underscored the ways in which SC's organizational and communication skills facilitated their work within the school district. A representative from a local clean air advocacy group underscored these skills, reporting, "[Prior to her coming], we were reaching one or two schools at a time. Now I have a central person I can work with to access so many other schools."

Principals also described the value of the network the SC created within the district as she established consistent and transparent points of contact and responsibility for various green initiatives. According to one middle school principal, "It's as though she said, 'Look, I'm the bridge.' And, you feel like you've been pounding your head against the wall and, suddenly, there's a door there and a gatekeeper you know."

Engaging Green benefited from the SC's ever expanding network of influence, at the same time the project strengthened the network on behalf of other district sustainability-related initiatives and reforms. For example, the SC invited local and state level officials to serve as judges for the Engaging Green Showcase, including representatives from more than ten high profile community partners including state and federal agencies, non-profit organizations, and businesses.

### *District Established a District Level Green Team*

In early fall of her first year, the SC held a full-day working session with the directors of key district-level departments, including accounting, budgeting, nutrition services, transportation, accountability, curriculum, purchasing, utilities, operations, planning, and construction to discuss the preliminary analysis of district policies and practices. At the session, the SC facilitated discussion regarding district-wide sustainability goals and the short and long term changes necessary to accomplish those goals. Functioning as the District Green Team, these individuals assumed responsibility for various tasks and a corresponding timeline.

The district-level Green Team's top-down support made all the difference in nurturing cultural change within the school district. As an internal change agent, situated in the middle of the district hierarchy, the SC lacked the positional authority to mandate change. Support from the highest levels of the organization paved the way for her work, accelerating changes in attitudes, norms, and practices district wide. One member of the district Green Team observed the progress in accomplishing this system-wide change.

> I think at first the immediate reaction is, oh, it is one more thing I have to do, but, as the work continues and we begin to shift the culture, the whole process becomes imbedded in the work we do. For one small example, recently we hosted a conference with 250 people, and we only used one ream of paper.

The director of transportation for the district offered his own evidence of change.

> In transportation, we have taken an active role in reconfiguring the way we do business, adding particulate traps for our busses to reduce nitrous oxide. We also use 16 compressed natural gas busses that further reduce nitrous oxide omissions, taking advantage of bus replacements through state funds. I remember in the early days coming to work early in the morning and there was this big cloud of blue smoke that hung over the transportation yard. You do not see that cloud anymore.

The Sustainable Grounds Manager, a newly created position on the operations side of the district observed, "It's become a requirement. The expectation is that we're all thinking about this together and we're all coming up with solutions together."

### *Engaging Green Facilities Projects Implemented*

The SC recognized the importance of "gaining the trust of principals and teachers." She focused on communication and relationship building on the one hand,

and effective program implementation on the other. As she nurtured important connections between and among people, at the same time she demonstrated her capacity to deliver on her commitments, establishing herself as a trustworthy leader and colleague. She explained:

> A lot of the principals and teachers didn't believe the Engaging Green funds would actually be used for their schools. The more facilities projects I was able to implement, the more people believed the district actually intended to follow through and the money wasn't going to be swept into the general fund.

According to one elementary principal, "I just thought the Engaging Green was brilliant because it is project-based learning. It is where we want to go with Common Core, teaching science and math, *and* it is relevant, authentic learning." It should be noted that at the close of the SC's third year of employment with the GSSD, 31 separate Engaging Green projects had been completed, 13 projects were currently underway, and an additional 20 projects were forthcoming, with Engaging Green continuing today.

### *District Drafted a Sustainability-focused Facility Master Plan and Passed Bond Issues Totaling $440 Million*

In spring of the second year, the district developed a sustainable facilities master plan as the over-arching document that would guide all capital projects. The Assistant Superintendent for Facilities, a newly created position, explained:

> The master plan became a living document, identifying key objectives, according to the six CHPS (Collaboration for High Performing Schools) criteria (i.e., addresses integration, indoor environmental quality, energy, water, site, materials & waste management, and operations). This will be my primary document moving forward, making sure that the projects we do are implemented according to a green program.

By November of the same year, the district passed two major bond issues, totaling $440 million. Together, these bond programs would continue to cultivate the conditions within which sustainability-focused reforms would happen and novelty would emerge in bricks and mortar improvements, management practices, community partnerships, and curricular innovations. In fact, the bond dollars would also continue to support Engaging Green for the next five years. One teacher shared their elementary school's plan for future involvement in Engaging Green.

> With the Common Core we can break down the Engaging Green inquiry so each classroom will be assigned one area of investigation. It will be a year-long project, with each teacher taking a part. They will teach more

> deeply, and all the students will be engaged in some aspect of the audit, writing and applying math skills where necessary. Then, as a school, we'll decide on the one we think is best to submit for the Engaging Green presentation.

According to the GSSD superintendent:

> Now there is bond money that will continue to support Engaging Green for the next five years. That is important. We just raised expectations. It is like sunshine – the best disinfectant. We have put it out there, and now people expect it. It is hard to argue with that, particularly when your logo is green.

## Discussion and Implications

Close consideration of the various critical events reveals key reform champions and implementers who operated at various levels of the system, from the classroom, to the school, to the district, to city government, and out into the community at large. As organizational conditions changed, including the hiring of a new superintendent and the mayor's launch of a citywide greening initiative, school district leaders responded creatively by seeking an internal change agent to help facilitate their own district-focused reforms. In addition, Engaging Green emerged as a nexus for energy, placing students at the center and welding commitment to these sustainability-related reforms.

For the first time, one individual (the SC) assumed responsibility as point person for all things green. Interview participants underscored again and again the importance of having such a person who connected participants with one another. The SC helped individuals understand how their various role(s) connected, or could connect, within and across initiatives. Through these person-to-person connections, participants learned they were contributing to something larger than themselves, something that had the potential to improve the lives of students and the community.

District leaders established the "unambiguous purpose of achieving sustainability sometime in the future," and "rewarded ongoing learning and innovation to continually increase individual, team, and organizational understanding and knowledge" (Doppelt, 2010, p. 102). Through the efforts of the SC, participants learned the history of the district's greening efforts and came to better understand the social/cultural dynamics of their own community and school system. District and school leaders utilized this growing knowledge as a foundation for the establishment of new sustainability-focused values, norms, and symbols. In addition, the ongoing professional development, provided to the SC by the CSL, and imparted to key participants, helped the district build capacity for systems thinking and strategic leadership.

With Sweeney's training in systems thinking in mind, the SC facilitated novel interactions across the school district and broader community in order to broaden and deepen sustainability efforts.

Across the initial three years of her tenure with GSSD, the SC worked to ensure that the district's sustainability-related goals remained visible, modeled commitment to these goals, and persisted. She intentionally worked to connect elements of the system already working for sustainability while also introducing sustainability concepts to those unfamiliar with them. And, day-to-day, the school district leaders, building level staff, and the SC demonstrated how their actions resulted in value-added and cost saving outcomes, for the district, school sites, and especially students. They built on the ideas, expertise, and willingness already present and created new opportunities that provided the greatest leverage for facilitating district-wide change.

### *Building Diverse Systems of Networks and Relationships to Facilitate Healthy Organizational Change*

As stated earlier, the study of living systems reveals that systems are healthy to the degree they develop diverse communities, set within dense networks of relationships, to withstand and recover from disturbances. Wheatley and Kellner-Rogers suggest we "create better health in a living system, [when we] connect it to more of itself . . . A failing system needs to talk to itself, especially those it didn't know were even part of itself" (1996, p. 7). The SC utilized all the district's human capital at her disposal, just as she invented new resources in the form of community partnerships. She identified key advocates at each school site who would act as the point of contact for all things green, and called upon key district leaders to assist in various initiatives. She coaxed, cajoled, coached, and rewarded participation.

Large urban school districts are complex bureaucracies with traditional hierarchies and typical silos. Quite relevant to sustainability-related reforms is the typical division between the curriculum and maintenance and operations. Many green school initiatives address building improvements and sit squarely in the facilities department. However, green building projects have the potential to serve as powerful learning opportunities for students at all levels.

Across GSSD, there were substantial pockets of existing expertise related to sustainability and green school programs. These pockets included individual teachers, small teams, and larger programs. The SC worked to identify and link these pockets of expertise. As the SC tapped existing networks of formal and informal subgroups, these networks facilitated the SC's uptake and distribution of information and expertise (Daly, 2010). "Subgroups within an organization . . . form some of the basic relational structures through which resources [expertise, knowledge, skills] may flow and, as such, may either support or constrain efforts at change" (Daly, 2010, p. 266). Daly (2010) elaborated:

> Change strategies, no matter how well-thought-out or useful, are more likely to be adopted from a trusted colleague than from an unfamiliar expert. This implies that successful change requires not only attending to the important formal structures, but also to the informal networks of social relations that create webs of understanding, influence, and knowledge prior to, during, and after the implementation of a change strategy.
> *(p. 2)*

### *Allowing for Self-organizing, Self-regenerating Systems to Facilitate Healthy Organizational Change*

According to Doppelt (2010), sustainability initiatives succeed to the degree that sustainability-related thinking habits and behaviors are incorporated into the everyday operating procedures and culture of an organization. As participants begin to value new ways of being – such as caring for the environment, organizational members, and the health of the community – old behaviors grow inconsistent with these new values and may be deemed inappropriate. Likewise, as organizational leaders communicate unambiguous messages regarding these new values, at the same time they reward learning and innovation in support of these values, they increase individual, team, and organizational understanding. Finally, to the degree leaders create the conditions for teams and networks to self-organize, novel solutions and adaptations emerge.

The SC held a non-supervisory position within the district; she did not have positional authority that might have allowed her to mandate changes. Thus, her change strategies relied primarily upon her ability to leverage the nascent greening initiatives underway in the district, mining the information and energy contained within these activities as fuel to reproduce them. Because her entire array of responsibilities centered on sustainability, she continuously scanned the school district, and the community beyond, for related activities and connected various isolated pockets of innovative work. Her capacity for systems thinking helped her track and grow diverse networks of relationships for sustainability. She celebrated and supported already existing efforts and forged the necessary relationships for initiating new work. And, as the district reforms gained notoriety and strength, the system realized a level of "self-maintenance via a mechanism of self-regeneration from within" (Capra & Luigi, 2014, p. 131). Evidence of this self-generation includes the institutionalization of the SC as a permanent district position, the passage of the combined bond issues totaling $440 million, the hiring of an Assistant Superintendent for Facilities to shepherd the Sustainability-focused Facility Master Plan, and the ongoing funding of Engaging Green.

## Conclusion

Through identification and discussion of critical incidents, the research participants provided feedback regarding the effectiveness of SC and district leadership, the challenges leaders and school district members faced in institutionalizing environmentally sound practices, and the various strategies employed to overcome reform-related obstacles encountered at the individual, school, and/or district level. The data reveal noteworthy changes in the broad array of sustainable practices related to energy and water conservation, waste reduction, healthy and locally grown lunch programs, and greener purchasing practices. In addition, students were afforded integrated, place- and problem-based learning opportunities, engaging them in an energy auditing process at their schools. These audits resulted in student recommendations for needed green-related facilities up-grades funded through local bond monies.

Educational leadership for sustainability constitutes a vehicle for 21st century best practice in education. Systems thinking is an invaluable capacity of this work, both for understanding our interdependence with all life systems and for informing strategic leadership for change. This study provides important insight into the ways a large urban school district applied living systems understandings of organizational change, opening their system to the energy and expertise provided by a non-profit partner organization as a means to invite creative solutions and adaptations and build a diverse network of support for sustainability-related building, management, and curricular and instructional practices on behalf of their students and families, their staff, and their communities at large.

## References

Bettencourt, L. M. A., & Kaur, J. (2011). Evolution and structure of sustainability science. *Proceedings of the National Academy of Sciences*, *108*, 19540–19545.

Bernstein, T. (2003). *Building healthy, high performance schools*. Washington, DC: Environmental Law Institute. Retrieved from Environmental Law Institute website: https://www.eli.org/research-report/building-healthy-high-performance-schools-review-selected-state-local-initiatives

Birney, A., & Reed, J. (2009). *Sustainability and renewal: Findings from the leading sustainable schools research project*. Nottingham: National College for Leadership of Schools and Children's Services. Retrieved from Institute of Education, University of London website: http://dera.ioe.ac.uk/2061/

Blank, M. J., Jacobson, R., & Melaville, A. (2012). *Achieving results through community school partnerships: How district and community leaders are building effective, sustainable relationships*. Washington, DC: Center for American Progress.

Bryk, A. S., Gomez, L. M., Grunow, A., & LeMahieu, P. G. (2015). *Learning to improve: How America's schools can get better at getting better*. Boston, MA: Harvard Education Press.

Capra, F. (1996). *The web of life*. New York: Anchor Books.

Capra, F. (2002). *Hidden connections*. New York: Doubleday.

Capra, F. (2009). *The new facts of life*. Center for Ecoliteracy. Retrieved from Center for Ecoliteracy website: http://www.ecoliteracy.org/essays/new-facts-life

Capra, F. (2010). *Life and leadership for a sustainable community*. Retrieved from http://www.ecoliteracy.org/article/life-and-leadership-sustainable-community

Capra, F., & Luigi, L. P. (2014). *The systems view of life: A unifying vision*. Cambridge: Cambridge University Press.

Cassell, J. A., & Nelson, T. (2010). Visions lost and dreams forgotten: Environmental education, systems thinking, and possible futures in American public schools. *Teacher Education Quarterly, 37*, 179–197.

Coburn, C., & Russell, L. R. (2008). District policy and teachers' social networks. *Educational Evaluation and Policy Analysis, 30*, 203–235.

Daly, A. J. (2009). Rigid response in an age of accountability. *Educational Administration Quarterly, 45*(2), 168–216.

Daly, A. J. (Ed.) (2010). *Social network theory and educational change*. Boston: Harvard Education Press.

Daly, A. J., & Finnigan, K. S. (2010). A bridge between two worlds: Understanding network structure to understand change strategy. *Journal of Educational Change, 11*, 111–138.

Daly, A. J., Liou, Y. H., Tran, N. A., Cornelissen, F., & Park, V. (2013). The rise of neurotics: Social networks, leadership, and efficacy in district reform. *Educational Administration Quarterly, 50*(2), 233–278.

Denzin, N. K. (1970). *The research act: A theoretical introduction to sociological methods*. Chicago: Aldine Publishing Company.

Doppelt, B. (2010). *Leading change toward sustainability: A change-management guide for business, government, and civil society*. Sheffield, UK: Greenleaf Publishing.

Edwards, B. W. (2006). Environmental design and educational performance. *Research in Education, 76*, 14–32.

Fook, J. (2002). Theorizing from practice: Toward an inclusive approach for work research. *Qualitative Social Work, 1*(1), 79–95.

Goleman, D. (2009). *Ecological intelligence: How knowing the hidden impacts of what we buy can change everything*. New York: Broadway Books.

Goleman, D., & Senge, P. (2014). *The triple focus: A new approach to education*. Florence, MA: More Than Sound.

Gordon, D. E. (2010). *Green schools as high performance learning facilities*. Washington, DC: National Clearinghouse for Educational Facilities, National Institute of Building Sciences.

Hawken, P. (2007). *Blessed Unrest: How the largest movement in the world came into being and why no one saw it coming*. New York: Penguin Group.

Heers, M., Van Klaveren, C., Groot, W., & Maassen van den Brink, H. (2016). Community schools: What we know and what we need to know. *Review of Educational Research, 86*, 1016–1051.

Henderson, K., & Tilbury, D. (2004). *Whole-school approaches to sustainability: An international review of whole-school sustainability programs*. Canberra, Australia: Australian Research Institute in Education for Sustainability.

Higgs, A. L. & McMillan, V. M. (2006). Teaching through modeling: Four schools' experiences in sustainability education, *The Journal of Environmental Education, 38*(1), 39–53.

Huckle, J., & Wals, A. E. (2015). The UN Decade of Education for Sustainable Development: Business as usual in the end. *Environmental Education Research, 21*(3), 491–505.

Kvale, S. (1996). *Interviews: An introduction to qualitative research interviewing*. Thousand Oaks, CA: Sage.

Kensler, L. A. W., & Uline, C. L. (2015). The transformation of a school district from energy hog to energy star. In Gross, S. J. and Shapiro, J. P. (Eds.), *Democratic ethical educational leadership: Reclaiming school reform*, pp. 54–58. New York: Routledge.

Kensler, L. A. W., & Uline, C. L. (2017). *Leadership for green schools: Sustainability for our children, our communities, and our planet*. New York: Routledge.

Kong, S. Y., Rao, S. P., Abdul-Rahman, H., & Wang, C. (2014). School as 3-D textbook for environmental education: Design model transforming physical environment to knowledge transmission instrument. *Asia-Pacific Journal of Teacher Education, 23*(1), 1–15.

Merriam, S. B. (1998). *Qualitative research and case study applications in education*. San Francisco, CA: Jossey-Bass.

Morgan, K., & Sonnino, R. (2008). *The school food revolution: Public food and the challenge of sustainable development*. London: Earthscan.

NCES (2013). Fast facts, available online: https://nces.ed.gov/fastfacts/display.asp?id=55

Nolet, V. (2015). *Educating for sustainability*. New York: Routledge.

Nolet, V. (2009). Preparing sustainability-literate teachers. *Teachers College Record, 111*, 409–442.

NWF (2016). Calling all eco-schools. Retrieved from http://www.nwf.org/Eco-Schools-USA.aspx

O'Conner, J., & McDermott, I. (1997). *The art of systems thinking: Revolutionary techniques to transform your business and your life*. New York: HarperCollins.

Patton, M. Q. (1990). *Qualitative evaluation and research methods* (2nd ed.). Newbury Park, CA: Sage.

Peterson, G., Allen, C. R., & Holling, C. S. (1998). Ecological resilience, biodiversity, and scale. *Ecosystems, 1*(1), 6–18.

Query, J. L., Kreps, G. L., Arneson, P., & Caso, N. S. (2001). Toward helping organizations manage interaction: The theoretical and pragmatic merits of the critical incident technique. In S. L. Herndon & G. L. Kreps (Eds.), *Qualitative research: Applications in organizational life* (pp. 91–119). Crossbill, NJ: Hampton Press.

Rauch, F. (2002). The potential of education for sustainable development for reform in schools. *Environmental Education Research, 8*(1), 43–51.

Reina, V. R., Buffel, T., Kindekens, A., De Backer, F., Peeters, J., & Lombaerts, K. (2014). Enhancing engagement through a community school approach as the key to increase academic achievement. *Procedia - Social and Behavioral Sciences, 116*, 2078–2084.

Satullo, S. K. (2012, January 26). Broughal middle school's after-school programs thriving. *Lehigh Valley Live*. Retrieved from http://www.lehighvalleylive.com/bethlehem/index.ssf/2012/01/post_187.html

Scott, W. (2013). Developing the sustainable school: Thinking the issues through. *Curriculum Journal, 24*(2), 181–205.

Selby, D. (2000). A darker shade of green: The importance of ecological thinking in global education and school reform. *Theory into Practice, 39*, 88–96.

Senge, P. M. (1990). *The fifth discipline: The art & practice of the learning organization*. New York: Doubleday Business.

Stake, R. (1995). *The art of case study research*. Thousand Oaks, CA: Sage.

Stone, M. K. (2009). *Smart by nature: Schooling for sustainability*. Healdsburg, CA: Watershed Media.

Stone, M. K., & Barlow, Z. (Eds.). (2005). *Ecological literacy: Educating our children for a sustainable world*. San Francisco: Sierra Club Books.

Sweeney, L. B. (1995). *Systems thinking playbook*. White River Junction, VT: Chelsea Green Publishing.

Sweeney, L. B. (2008). *Living stories about living systems*. White River Junction, VT: Chelsea Green Publishing.

Taylor, A. (2009). *Linking architecture and education: Sustainable design of learning environments*. Albuquerque: University of New Mexico Press.
US Green Building Council (2017). Personal communication.
Wheatley, M. J. (1999). *Leadership and the new science: Discovering order in a chaotic world* (3rd ed.). San Francisco: Berrett-Koehler.
Wheatley, M. J., & Kellner-Rogers, M. (1996). *A simpler way*. San Francisco: Berrett-Koehler.
Wielkiewicz, R. M., & Stelzner, S. P. (2005). An ecological perspective on leadership theory, research and practice. *Review of General Psychology*, *9*(4), 326–341.
Williams, A., Kennedy, S. P., Philipp, F., & Whiteman, G. (2017). Systems Thinking: A review of sustainability management research. *Journal of Cleaner Production*, *148*, 866–881.
Wu, Z. (2002). Green schools in China. *The Journal of Environmental Education*, *34*(1), 21–25.

# 7

# "HOLISTIC ENGAGEMENT"

## Framing Theory, Strategic Communication, and the Superintendency

*James Coviello and David E. DeMatthews*

Despite years of reform and the tireless efforts of teachers, principals, and other key stakeholders, many schools and districts remain places of stubborn inequality. A significant body of evidence has revealed the persistence of inequitable policies and decision-making for historically disadvantaged students, evidenced in unequal achievement levels, school funding inequities, unequal program offerings and teacher quality, and harsh discipline policies and practices (Adamson & Darling-Hammond, 2012; Darling-Hammond, 2007; Gregory, Skiba, & Noguera, 2010). Although multiple factors contribute to this state of affairs, researchers have long noted the important role of educational leaders – like district superintendents and school principals – in altering the status quo and creating equitable change (Brooks & Miles, 2008; Brown, 2004; Furman, 2012; Theoharis, 2007).

Both superintendents and principals regularly make important decisions on policy and practice (Coburn, 2006) and their actions can have important consequences for reform efforts focusing on historically marginalized student populations. Numerous studies have documented and described the actions of social justice leaders – usually school principals – who put issues of equity central to their leadership efforts (DeMatthews & Mawhinney, 2014; Furman, 2012; Green & Dantley, 2013; Rivera-McCutchen & Watson, 2014). Many researchers of social justice leadership conclude their studies with a recognition that the challenges confronting schools and marginalized students are significant, deeply entrenched, and often emerge outside of the school and the principal's locus of control (Berkovich, 2014; Green & Gooden, 2014).

In addition to considerations of social justice leadership, scholars have also studied the implications of policy change and policy implementation on creating equitable change. Researchers have noted the role that educational leaders play in the adoption of equitable policy and the "translation" of these policies into practice (DeMatthews, 2015; DeMatthews & Mawhinney, 2014). Multiple studies have examined policy creation and implementation as a social activity, whereby members of an organization "make sense" of policy changes through personal relationships and ongoing social interactions (Coburn, 2006; Weber & Glynn, 2006; Weick, Sutcliffe, & Obstfeld, 2005). Coburn (2006) has emphasized the importance of "problem framing" as part of the process of creating and translating policy into practice and notes that this is a key element of educational leadership. Leadership for equity on both the school-site and district level must therefore take into account the importance of sensemaking and framing processes when attempting to alter unjust practices and policies.

To be certain, principals deserve attention for their role in influencing instruction and building-level decision-making. But despite their important role, less attention has been paid to the influence of superintendents in leading for equitable change (DeMatthews, Izquierdo, & Knight, 2017). As the day-to-day leaders of a school district, superintendents oversee budgetary and other district operations, help give practical meaning to federal and state policy, and set local district policy for the schools under their charge. Superintendents therefore work in a unique and important leadership space that exists between federal, state, local governments and the principals and teachers that educate students (Woulfin, Donaldson, & Gonzales, 2016).

In their attempts to change policy and practice in their districts, superintendents also have an important role in "framing" – or strategically communicating – policy messages to different stakeholder groups, including the elected school board, business and community leaders, parents, principals, and teachers (Coburn, 2006; Park, Daly, & Guerra, 2013; Woulfin et al., 2016). In this context, a superintendent seems to be in a strong position to shape and implement district policies that might allow schools to better serve groups of historically marginalized students and interrupt the status quo. However, the power of a superintendent to create change is restrained and attenuated by a complex landscape of constituencies and stakeholder groups (Kowalski, 2006; Kowalski et al., 2011). In short, if a superintendent perceives a program or policy to be unjust, it may be difficult, impossible, or politically unadvisable to unilaterally change it and expect this alteration to be durable. A superintendent may attempt to strategically frame socially-just policy changes in order to achieve authentic buy-in and longer-lasting results.

This kind of strategic thinking – in this instance a case of strategic communication – is rooted in holistic conceptions of district leadership. Rather than a transactional view of community or parent engagement like the ubiquitous "coffee with the principal" (Auerbach, 2010), framing processes happen on more of a "30,000-foot view," seeing the district as an interconnected web of complex social interactions comprised of individuals with their own agency and interests. Focusing solely on influencing the sense-making of what is perceived as the group with the most interest in a policy change – as with a teachers' union and a union contract – ignores the sense-making processes that are concurrently taking place with other stakeholders in the district. Framing the need for a policy change to one group and achieving buy-in will result in only short-term success if other stakeholder groups are ignored. A holistic view of district leadership sees efforts at engagement and communication not just as individual issues to address with disparate groups, but rather as a broad-based strategic effort for durable change.

Drawing upon historical analysis of the role of the superintendent, framing and social justice theories, we explore how one superintendent in a district in the southwestern U.S. led for equity by utilizing strategic communication to legitimize and gain support for equity-based policy change. This study therefore seeks to answer the following research questions: (a) how did the superintendent frame equity-based policy choices; (b) what is the substance (or content) of those frames; and (c) what factors did the superintendent consider when creating those frames?

What follows is an overview of the literature on the historical evolution of the role of the superintendent, social justice leadership, and sense-making and framing theories. Following the literature review is a discussion of the methodology used in this study. Next is a presentation of findings, including the major themes that emerged through an analysis of the data. Finally, the chapter ends with a discussion of the superintendent's role in framing an equitable policy change and will include a basic model for "holistic engagement" that recognizes the utility of broad and deep strategic communication strategies to create durable acceptance of equity-based policy changes.

## Literature Review

This review of literature will focus on several main strands: the evolving historical role of the superintendent, particularly the most recent role of "communicator"; theories of policy creation, interpretation and implementation utilizing sense-making, sense-giving, and framing; and social justice leadership, including calls for advocacy leadership intended to increase community capacity and stakeholder involvement.

### *The Evolving Role of the Superintendent*

The evolution of the role of the superintendent has been studied discursively by multiple scholars, most notably Callahan (1966) and Kowalski (2001). Callahan (1966) observed four role conceptualizations: teacher-scholar, manager, democratic-leader, and applied social scientist. More recently, Kowalski (2001) added the role conception of "communicator." Both researchers noted that as the role of the superintendent changed over time, none of the expectations of superintendent leadership was jettisoned. In fact, each role was added to the previous layers of leadership and responsibility that the superintendent was expected to fill. A brief historical summary of these role conceptions follows below.

Once initially a clerical position, the role of the superintendent changed rapidly as districts' enrollments continued to grow and the antebellum Common School Movement, led by Horace Mann, took hold. The large scale of instruction required increased standardization of curriculum and the supervision of vast numbers of teachers. Superintendents were tasked with filling this role and became "teacher-scholars," (Bjork, Kowalski, & Browne-Ferrigno, 2014; Kowalski, 2006) in charge of what would might now be termed "instructional leadership." However, the conception of the superintendent as teacher-scholar, though never completely abandoned, began to diminish after the beginning of the twentieth century (Bjork et al., 2014). The role of the superintendent continued to evolve during the Progressive Era, a time period that saw a dramatic economic shift as the U.S. economy was changing from agrarian to industrial. Business and industrial leaders began to apply modern scientific management theories to increase the efficient operation of their businesses. Like executives leading modern corporations, superintendents were expected to manage a budget, oversee personnel, and maintain facilities. Their primary role shifted towards becoming "managers" as their previous role as instructional leader was pushed to secondary status (Bjork et al., 2014; Glass, Bjork, & Brunner, 2000; Kowalski, 2005).

After the Great Depression, the role of the superintendent as "manager" waned and in its place grew a greater emphasis on the political aspects of the position, whereby the superintendent would lobby and compete with other elements of local government for state funding, get the local community on-board with education initiatives, and navigate the political demands of various stakeholder groups (Bjork et al., 2014; Bjork & Lindle, 2001; Kowalski, 2005). The superintendent would also be expected to participate in and foster local democratic control of schools and push back against the past movement towards corporate-style management models (Bjork et al., 2014). This perception of the superintendent as "democratic leader" eventually fell out of favor and was criticized as being too detached from the needs of actual practice (Kowalski, 2005). In the 1950s and 1960s, social science research critical of the inadequacy of large institutions like public schools became more well-known and led to calls for superintendents to take on the role of "applied social scientist" (Bjork et al., 2014; Kowalski, 2005). The idea behind this role was for the superintendent to use and apply scientific theories and methods to understand the problems in their district and help to solve them.

The most recent role, that of "communicator," meant that the superintendent is expected to build coalitions and sustain change through relationship building and cultural change within their organizations (Kowalski, 2005). As a communicator, a superintendent leads an organization through clear and consistent communication to stakeholder groups, both internal and external. This role has become so significant that, rather than a simple skill that is needed to lead effectively, it has become an indispensable superintendent role that stands on its own (Kowalski, 2005). The position of "communicator-in-chief" stems from the belief that school improvement can only take root at the local level through superintendents taking an important role in leveraging change through internal and external educational stakeholders. Within this role conception, superintendents will simply not be effective without a consistent ability to discursively signal their reform intentions and attain buy-in from varied constituencies. This most recent role of "communicator," as it relates to policy framing and advocacy leadership, will be discussed in further depth below.

## *Policy Creation as Social Process*

Scholars have noted that policy creation and implementation are social activities, whereby members of an organization "make sense" of policy problems and create policy prescriptions through personal relationships and interactions (Weber & Glynn, 2006; Weick et al., 2005). An informal understanding of this sense-making process is the construction of an answer to the question, "What is going on?" Educational researchers have been interested in understanding the process by which principals and teachers make sense of and interpret policy reform and how that process influences their daily practice (Coburn & Talbert, 2006; Spillane, Reiser, & Reimer, 2002). Researchers have also noted the role that educational leaders play in the adoption of equitable policy and the "translation" of these policies into practice (DeMatthews, 2015; DeMatthews & Mawhinney, 2014).

During the policy process, members of a school or district engage in a process of selecting information from their environment, creating meaning with that information, and then taking action based on that meaning (Coburn, 2006). Sense-making is often a contested process that involves issues of power and status (Weick et al., 2005). Leaders in educational organizations "give sense" and are situated to do so because of their positional status within their organization (Coburn, 2006). In other words, not all individuals participate equally in the sense-making process within an organization. However, in a dialectic process, group members give sense to each other and also give sense back to organizational leaders (Coburn, 2006). Several factors influence sense-making and sense-giving for educational leaders, including the socio-economic status of district and school members, including students, parents, teachers, staff, and community members (Evans, 2007; Spillane, Diamond, Burch, Hallett, Jita, & Zoltners, 2002).

## *Framing Theory and Strategic Communication*

Complementary to theories of sense-giving are studies involving framing analysis. Framing involves deliberate attempts to influence the process of sense-making for others and to inspire action. This is understandable given the origin of framing theories in social movement theories (Benford & Snow, 2000). Framing analysis is usually examined in "core framing tasks": diagnostic, prognostic, and motivational (Benford & Snow, 2000). Diagnostic framing involves the identification of the issue or problem that needs to be resolved. Prognostic framing is the promotion of a proposed course of action to resolve the issue or problem. Motivational framing is the creation and promotion of reason for taking action. Though presented here in a specific order, framing does not necessarily need to occur as such (Park et al., 2013).

Some recent research has focused on the important role that superintendents play in "framing" – or strategically communicating – policy messages to different stakeholder groups, including the elected school board, business and community leaders, parents, principals, and teachers (Coburn, 2006; Park et al., 2013; Woulfin et al., 2016). Framing theory can therefore be useful as a framework for understanding educational leadership and educational organizations. Leaders, in positions of formal authority, play an important – though certainly not an exclusive role – in influencing the process of sense-making for others in an organization (Park et al., 2013; Woulfin et al., 2016). They are also leaders in a politicized environment, where their place in the organization makes them "positioned to frame, direct, support, constrain, or transform practitioners' sensemaking of reform" (Park et al., 2013, p. 668).

In the politicized policy context of schools, theories of framing are useful in understanding how leaders utilize strategic communication to legitimize and gain support for policies (especially equity-based policies) in their districts. Implementing policies with fidelity is an important goal for policy creators, and framing facilitates an understanding of the process that leaders undertake when they attempt to achieve that goal. Park et al. (2013) recognized the ability of framing to gain buy-in, writing that leadership around policy "is not only about managing technical expertise but also an inherently political process by which certain frames are valued, and as such, shape meaning and potentially outcomes" (p. 668). In the conflict-laden, democratic context of school districts, a superintendent – in assuming the recent discursive role of "communicator" – seems to be in a strong position to frame district policies, shape the sense of various stakeholders, and gain support for possibly contentious policies that focus on groups of historically marginalized students and interrupt the status quo.

## *Social Justice Leadership*

Educational researchers have long recognized the kind of leadership that focuses on historically marginalized students and attempts to interrupt the status quo: social justice leadership. Social justice is itself a contested and somewhat ambiguous term, with some scholars writing that social justice "encompasses a range of terms—some more powerful than others—such as equity, equality, inequality, equal opportunity, affirmative action, and most recently diversity" (Blackmore, 2009, p. 7). For the purposes of this study, we define social justice leadership as an orientation through which leaders recognize and attempt to address the structural inequities facing historically marginalized, underserved groups in schools (Dantley & Tillman, 2006; Furman, 2012). Social justice leadership recognizes that effective leadership requires more than one person in one school – it requires a systemic or ecological view of communities, schools, and districts. This kind of leadership prioritizes issues of equity or access for all students in their communities, including groups that may have been previously pushed to the margins – Black and Latino students, ELLs, LGBTQI, and students with disabilities.

To create this kind of systemic change, researchers have noted the need for school leaders to form lasting partnerships and have authentic collaboration with their surrounding communities – not just "coffee with the principal" (Auerbach, 2010). Instead of focusing on their obvious challenges, social justice leaders try to recognize the assets within their community and their school's distinctive capacity to be a driving force for revitalization and improvement. These leaders proactively seek out relationships with parents and community members, creating networks of allies and advocates in the process (DeMatthews, 2018). By doing so, leaders help create the capacity for sustainable, positive change in their schools and their communities – and not leverage parental involvement solely to improve test scores or other district priorities. Some researchers have called for school leaders to transform their practice into a form of community or "advocacy leadership," whereby they tap into the resources within a community to build the capacity for large-scale, sustained change (Anderson, 2009; Ishimaru, 2013). Rather than staying silent, social justice leaders find ways to advocate for the needs of their schools and communities.

In sum, superintendent leadership and district policies play a role in creating equitable change, yet this topic remains understudied. This review of literature focused on highlighting several factors related to understandings of superintendent leadership, including the importance of the superintendent on district performance and the evolving historical role of the superintendent. The most recent role iteration – "communicator" – is important and connects to theories of policy creation, interpretation, and implementation. In particular, framing analysis provides a lens to examine the role of superintendent as communicator when attempting to implement contentious policy changes. Theories of social justice leadership call for placing issues of equity at the center of leadership practices and call for advocacy and community capacity-building that will create large-scale support for long-lasting equitable and systemic change. A superintendent, acting as communicator, might use "framing" to build community capacity and stakeholder support for equity-based policies that social justice theorists say is required.

## Methodology

In the previous section, we weaved together multiple strands of literature on the superintendency, policy theories of sense-making and framing, and social justice leadership. We believe that these components are applicable to superintendents seeking practical avenues for creating more equitable and socially just school districts. Drawing on this literature, this qualitative case study examines the leadership of a superintendent in creating a policy change benefiting a historically marginalized population in his district – English Language Learners (ELLs). In this case, the superintendent, "Mr. Maldonado," attempted to expand a dual language program for ELLs from a small pilot program to a district-wide policy (pseudonyms were used for all participants and schools). We examine his attempts at strategically communicating his new policy direction as well as his specific actions, orientations, and beliefs that informed his framing choices.

We collected our primary data from semi-structured interviews with Mr. Maldonado and triangulated his responses with interviews with other key stakeholders, including a deputy superintendent, school principals, teachers, and parents. Four interviews were conducted with Mr. Maldonado, each lasting between one and two hours. Each interview was tape recorded and transcribed. Data was also derived from observations of school board meetings, meetings of a district task force on dual language, and professional development sessions. In total, over 100 hours of observation took place over the course of one year. Following each interview and observation, the researchers reviewed observation notes, linked any collected documents to the notes, and created a file for the observation. We chose these data sources based on similar studies that utilize framing and sense-making as well as our research questions, which aim to understand the lived-experience of superintendents when framing a policy change, rather than the effectiveness of that framing or of the policy itself. These three sources of data also allowed for an in-depth exploration of their framing experiences and how they made decisions about messaging and strategy for their communications with different stakeholder groups.

### *Data Analysis*

We used NVivo10 to organize, code, and analyze the collected data in multiple phases. First, we read and re-read each data source several times (interviews, observations, and documents), looking for broad themes and taking brief notes of these themes as well as marking key quotations. We recorded data by type and source and began with an initial coding phase that involved low-inference codes derived from our conceptual framework, specifically Benford and Snow's (2000) three-part conception of framing. This framework consists of three modes of framing: diagnostic, prognostic, and motivational. Codes were based on these modes and on emergent themes related to the language of the frames of leaders, the reasons they provided for framing a policy in the way that they did, and on themes related to changes to the sense and meaning making of constituent groups. We also maintained prolonged engagement at the research site which allowed us to build trust with the superintendent, deputy superintendent, and other participants (Miles & Huberman, 1994).

## Findings

We present our findings as they relate to the context of the district and the role of Superintendent Maldonado's leadership in implementing a dual language policy. First, we briefly describe the district context prior to Superintendent Maldonado's arrival. Second, we highlight his early steps to expand dual language from a small program to a district-wide policy. Third, we examine Maldonado's role as "communicator" at the district and community level and the specific district practices that contributed to the implementation of the policy. Finally, we conclude with a description of the opportunities for framing and broadening a message of equity that Maldonado missed in his leadership efforts.

### *A Context of Mistrust and Upheaval*

Mountain Hills Independent School District (MHISD) is a large urban school district located in the U.S. Southwest along the U.S.-Mexico border and includes some of the poorest zip codes in the nation. The district enrolls more than 60,000 students – approximately 80% of students are Hispanic, 12% White, 70% economically disadvantaged, and 25% ELL. The district had gone through several years of upheaval, including a criminal conviction for the previous superintendent based on charges relating to a scheme to artificially improve test scores by keeping low-performing students out of tested grade levels.

Twenty years before this scandal, a small cluster of schools in a wealthier area of the city had begun developing dual language programming independent of the district office. One of these schools had in fact received national attention for its advancement of dual language education and eventually partnered with a middle school and high school to continue the program for their students. However, because of their location in a more affluent area, these schools mostly served native English-speaking students from higher socio-economic backgrounds whose parents were interested in dual language as more of an enriching program choice. The vast majority of MHISD relied on a policy of early-exit bilingual education.

At the beginning of the 2013 school year, the MHISD school board appointed an attorney and businessman, Jose Maldonado, as superintendent. This choice was criticized, mostly due to his lack of school leadership experience. However, considering national and statewide trends, his background was notable: he is a Mexican-American, born in the area, and grew up in a Spanish-speaking household.

In December, the district board created a community task force to examine the possibility of adopting dual language. Two months later, the group recommended a district-wide adoption of a policy of dual language for all ELLs.

## *Communication and Framing Change*

Though the initial creation of a dual language program was driven by affluent parent groups, Maldonado and his assistant superintendent recognized the existence of a grassroots energy created from the reaction to the MHISD scandal. Maldonado attempted to harness this energy by framing a dual language policy change as a requisite response to the cheating scheme. He communicated that this injustice necessitated systemic and ideological change. When speaking about dual language policies, he noted his own personal link, stating: "I grew up as an English language learner. I know what it's like trying to learn English and what those experiences are like." One former dual language task force member and district parent spoke in terms that highlighted the need to fix previous maltreatment: "It's horrible what they did and it really set back the district. There's no trust, plus, the harm they did to so many kids. You can't just repair that damage overnight." Maldonado recognized that the situation dictated a need for the district to listen to parent concerns and put those concerns at the center of reform efforts. He explicitly said "I must be an advocate for our ELL's. They must be our priority now." When Maldonado communicated with stakeholders like the school board, community members, principals, and teachers, he continually retold how emergent bilinguals were historically marginalized in the state and in the district and must be offered a more ideal future. Maldonado and his assistant superintendent framed injustice as a cause for action, an opportunity for change, and a sense of purpose connected to their own isolation and embarrassment during their personal stories of public schooling.

Maldonado desired to speak as an advocate for parents, pushing them to be involved in shared governance at the district-level. At the same time, he strategically employed this advocacy ability by co-opting influential stakeholder voices to build support for a policy focused on increasing equity. He reflected, "It [dual language] was really a grassroots community effort. I think that there were a number of parents who started dual language in the nine schools and their primary concern was getting more middle schools." Maldonado saw that parents with high socio-economic status and accompanying political influence could be utilized to help drive momentum for dual language policies for emergent bilinguals in the district and combat potential resistance to these changes. These parents, initially supportive of dual language programming for their own children, were an important source of support for expanding the policy for the rest of the district.

Despite efforts to repair the damage inflicted from the cheating scandal, interviews of parents, teachers, and principals showed that trust in the district had not been restored. In observations, most stakeholders showed themselves to be broadly supportive of dual language policies but were less trusting that the policy could be successfully implemented. Many principals reported that their schools were not ready for the changeover to dual language and were concerned with possible negative short-term effects on standardized test scores. One principal asked, "Do they really want us to do this well? If so, wouldn't we have more training and more time?" Many teachers had similar feelings of unpreparedness. However, parents were generally supportive of the switch to a dual language policy for emergent bilinguals or as an option for native English-speaking students. Maldonado listened and publicly acknowledged these challenges were real, but argued that change should happen despite potential dips in test scores or self-doubts over preparedness. Privately, he acknowledged that it was important to ensure dual language was implemented and solidified given the potential for shifting beliefs with new school boards and fluctuating state policies.

### *Driving Change and Responding to Resistance*

MHISD had a short time between the task force recommendation and the next school year when dual language implementation in pre-k and kindergarten would begin district-wide. Recognizing the need, and given the complexity and the ideological shift required, Maldonado and his assistant superintendent initiated professional development for principals, teachers, and parents the day after approval. Maldonado recognized the symbolism of starting immediately: "The fact that we started immediately communicated to schools that this was serious, that dual was coming."

However, when 2014–2015 began, many teachers and principals were frustrated by the rapid implementation of dual language. Maldonado and his assistant superintendent sought advice from teachers, professors, and other stakeholders and ultimately agreed to create teacher committees and teacher roundtables around problematic issues. Committees developed materials and provided schools with guidance. One district administrator commented:

> The committees made this our program, not something from the outside . . . We created teacher committees to have conversations about different teacher needs. Teachers who had experience in dual were very helpful to these committees . . . They shared ideas and resources and helped to lower the frustration level . . . It also made the reform feel more authentic, like it wasn't just this top-down thing.

Despite this, committee members and facilitators reported how teachers came to meetings upset. A participant said, "There was tension in the roundtables. Teachers felt they weren't ready and they wanted training."

Maldonado recognized the value in giving teachers these opportunities, saying "They need to share their ideas and frustrations."

Maldonado also realized how some principals, especially in the first weeks of implementation, did not buy-in to the model or became passive resistors who did not publicly challenge dual language, but were slow to support it. His assistant superintendent highlighted the importance of hiring and retraining principals, while also recognizing that some principals would not fit: "As we are replacing principals [due to retirement, transfer to other district, removal], we are replacing them with new blood that will be supportive . . . we have a process and by the time we pick the person, we make pretty darn sure that person is supportive of dual language." The recommendation was made that an "exemplar" dual language principal would facilitate monthly dual language professional development sessions with principals. While buy-in for dual language was slow, we found that the use of an exemplar principal as a professional development resource was a strategic move by Maldonado, as he understood how to use these well-placed advocates to promote reform and mitigate resistance.

### *Missed Opportunities for Framing and Engagement*

Although Superintendent Maldonado made efforts to engage with a variety of stakeholder groups, these efforts were motivated at least in part from a transactional perspective and were thereby limited in scope. For example, Maldonado focused his strategic communication on what he thought were the groups most directly linked to the immediate establishment of a dual language policy and whose support he deemed most necessary: parents, teachers, and principals. However, other stakeholders – whose support for long-term acceptance of a dual language policy might be necessary – were not deemed a priority. This meant that potentially important opportunities were missed to frame this policy change and influence the sense-making of a broader base of the community. Most notably, a case could have been made to gain the support of local business leaders who draw many of their future employees from MHISD and might have been an important group with which to engage.

Additionally, Maldonado's efforts at framing the policy change to dual language was hampered by the speed with which he decided to make the change. By not providing the time to create broad and deep-seated support through a campaign of sustained engagement and communication, the long-term survival and implementation of this policy might be in question. Without this kind of support, a near-term change in the make-up of the school board might result in pressure to change the policy (or even to replace Maldonado) and a quick reversal of the decision. In addition, a lack of time to make his case to principals, teachers, and staff may lead to a superficial level of policy implementation and passive resistance to this change. However, these shortcomings do not diminish the important and intentional framing efforts of Superintendent Maldonado but are instead examples of the complexity of taking a broader and holistic approach to community engagement.

## Discussion

This study explored the experience of the MHISD superintendent hired in a scandal-plagued district and his efforts to build support and buy-in for an equity-based policy change. Few studies have examined how superintendents "frame" policy reform to help promote more equitable policies and protecting needed changes in their districts. This study highlights the importance of a broad conceptualization of strategic communication when pushing for equitable reform in schools. As evidenced in our findings, Superintendent Maldonado was able to utilize multiple approaches when advocating for policy changes, from referencing an emotional personal connection to language acquisition to communicating effectively with a powerful parent group. These findings correspond with discursive descriptions of the role of the superintendent as communicator (Kowalski, 2001) and theories of policy framing (Benford & Snow, 2000; Coburn, 2006) and social justice leadership that recommends making advocacy central to leadership efforts (DeMatthews, 2015; Ishimaru, 2013).

Maldonado recognized the importance of his role of communicator as evidenced by his recognition that change required communication through ongoing open forums. This advocacy work was not a function of typical power politics where interest groups and individuals battle for resources, but rather a process of social construction where the superintendent framed key policy problems that the district should address with specific policy solutions (Coburn, 2006). Rather than utilize authoritarian approaches or mandates in the name of social justice, Maldonado and his deputy created forums and professional development opportunities where teachers and principals across the district could pose questions, share strategies, and problem-solve the unique challenges. In part, this was because Maldonado was able to think on multiple layers about how change could be slow and frustrating for a family, a classroom teacher, or a principal. Throughout the course of the study, he continued to learn, reflect, and refine his approaches as well as his ongoing awareness that equity-oriented change was a bumpy road.

Although he had not intended to utilize a particular framing or sense-making method, Maldonado's communicative practices can be seen through the lens of Benford and Snow's (2000) three-part conception of framing. When communicating with various stakeholders in different settings, he used diagnostic, prognostic, and motivational framing in order to effectively advocate for the policy change to dual language. For instance, Maldonado framed the cause of the problem (diagnostic) when he spoke of how the scandal damaged the lives of marginalized ELL students and highlighted the existence of a historical legacy of national, state, and district ELL policies that had left ELL students underserved for years. When advocating for the solution to the problem of ELL injustice (prognostic), Maldonado cited the success of the current dual language pilot program and the support that the program – and its expansion – had from parents whose students were currently enrolled in the program. Lastly, Maldonado utilized several modes of motivational framing when pushing for a speedy policy change. In an attempt to allay concerns over the policy's implementation, he offered words of encouragement and support to principals and teachers. He was also able to speak of the recent support for the policy resulting from the work of the community task force. Maldonado was also able to leverage his own personal story of isolation and struggle as an English language learner when communicating the immediacy of this policy change. Maldonado made these efforts to discursively motivate and inspire others to act and buy-into the district's dual language policies.

Superintendent Maldonado's view of engagement and his efforts at framing the policy change for dual language remained limited and somewhat narrow. A dual language policy, with its focus on bilingualism and biculturalism, provides numerous opportunities for diagnostic, prognostic, and motivational framing messages that might have been effective with groups that Maldonado failed to meaningfully engage. For example, a policy change like dual language might be diagnostically framed for local business leaders by noting the economic struggles, as well as the economic opportunity, of a bicultural and bilingual region. An educational leader might prognostically frame the issue by highlighting the demonstrated and measurable effectiveness (and therefore resource efficiency) of dual language policies. A motivational framing for business leaders might include the message that current language policies are currently a misallocation of resources and a missed opportunity for "workforce development" needs. This framing could note the higher "return on investment" of dual language and the "asset" creation of a bicultural and bilingual workforce that would be key for the economic growth of the region. This kind of framing language would be a departure from, for instance, Superintendent Maldonado's motivational framing of the issue as "the right thing to do for ELLs."

The importance of "holistic engagement" is grounded in efforts to effectively and strategically engage with as broad a spectrum of stakeholders as possible, even with those that may not at first glance seem to be directly affected by a given policy change. This is rooted in conceptions of systems thinking, with its focus on, as Senge (2006) notes, "seeing wholes. It is a framework for seeing interrelationships rather than things, for seeing patterns of change rather than static 'snapshots'" (p. 68). Simple efforts at communication with one or two affected groups ignores the interconnection of interested parties and the need for achieving lasting buy-in. This systems view of district leadership and strategic communication is focused not on discrete and separate issues in need of attention, but on a strategic effort to create structural change that persists past the often short tenure of an individual leader.

It would be prudent to reiterate that an evaluation of the success of Superintendent Maldonado's framing efforts – as well as the success of the district's policy – were outside the scope of this study. We do not desire to ignore their effectiveness or downplay the role of other factors – like organizational structure, district culture, and other issues of local context – in the successful adoption of this policy. Nevertheless, superintendent leadership actions in communicating and framing policy changes are key to an understanding of social justice leadership, advocacy, and community engagement.

## Conclusion

This study raises a number of questions that might ground further inquiry into superintendent leadership, policy implementation, and equitable change. First, how might superintendents broaden and deepen their role as "communicators"? What can they do to be more "holistic" in their engagement efforts and help leverage change in their districts? MHISD represented a case with significant pre-existing channels for engagement and communication and multiple stakeholder groups open to change due to an ongoing scandal. This study is limited by drawing on only one superintendent's experience in one particular district's context. Future research might explore superintendents' efforts to think and speak "bigger" and more broadly by expanding the number of stakeholder groups that superintendents engage in framing efforts. Superintendents might also expand and deepen their consideration of alternative framing messages for these groups.

For example, Maldonado recognized the value of opening up new, alternative venues for communication and framing through the establishment teacher committees and roundtables. Further studies might include superintendents' efforts in engaging a broader cross-section of stakeholder groups. The business community – often a powerful interest group – might attempt to influence policy creation at the district level and thus require alternative framing message strategies. Equity issues like ELL language policy often have a larger state, regional, or national political context, so future research might include attempts to include these factors in strategic communication efforts. Additionally, our study does little to understand or examine how superintendents with backgrounds different than the majority of their district attempt to frame equity-based policy changes. Maldonado was uniquely positioned to leverage his upbringing and experience to help with his efforts at communicating and engaging with district stakeholders, so future studies of social-justice oriented superintendents should include cases of varying background.

This study highlights the important role of the superintendent in engaging in strategic communication to support the creation and implementation of an equity-based policy. While MHISD provided a unique case for examining the particular contextual factors that might influence stakeholder engagement, the findings of this study highlight relevant lessons for superintendents interested in creating equitable change in their districts. First, superintendents should take a holistic view of communication and stakeholder engagement when attempting to create equitable policy change. Traditional forms of community engagement and communication focus on the short-term, are often tailored to a narrow view of student achievement (test scores), and include small slices of stakeholders (parents). A holistic orientation of community engagement and communication is broad, deep, and durable. Superintendents seeking holistic engagement see community and stakeholder interactions not as a method to increase test scores or keep interest groups happy, but as a practical tool to help achieve socially just ends. Superintendent Maldonado had personal knowledge of the marginalization and struggle that faces ELL students, but he recognized that his own desire for social justice was insufficient without broad-based community support and educator buy-in. He spoke to differing groups about the issue at hand and attempted to frame a message to try to get their buy-in. He also attempted to include any and all important stakeholders and prolonged his engagement with said groups to get and keep their support.

At the same time, Maldonado was savvy and strategic in his communication, framing his message to particular stakeholder audiences. This approach requires leaders to attempt to understand their communities – either through individual background or other personal connection and experience. Understanding of this kind will yield knowledge of the language and values of the group that is being addressed and will allow a superintendent to utilize this language when strategically framing a policy choice. Superintendents should thereby take into consideration the perspective of multiple stakeholder groups in the community and expand their voices for multiple audiences and on multiple levels of policy – including local, state, and national. A holistic view of engagement recognizes the value of all of these groups and all of these approaches. Without such a recognition, superintendents will struggle to frame their communication to reach these different groups and be constrained in their efforts to implement more equitable policies in their districts and challenge the status quo.

## References

Adamson F., & Darling-Hammond L. (2012). Funding disparities and the inequitable distribution of teachers: Evaluating sources and solutions. *Education Policy Analysis Archives, 20*.

Anderson, G. L. (2009). *Advocacy leadership: Toward a post-reform agenda in education*. New York: Routledge.

Auerbach, S. (2010). Beyond coffee with the principal: Toward leadership for authentic school–family partnerships. *Journal of School Leadership*, *20*(6), 728–757.

Benford, R. D., & Snow, D. A. (2000). Framing processes and social movements: An overview and assessment. *Annual Review of Sociology*, *26*(1), 611–639.

Berkovich, I. (2014). A socio-ecological framework of social justice leadership in education. *Journal of Educational Administration, 52*(3), 282–309.

Bjork, L. G., Kowalski, T. J., & Browne-Ferrigno, T. (2014). The school district superintendent in the United States of America. In A. E. Nir (Ed.), *The educational superintendent: Between trust and regulation—an international perspective* (pp. 17–38). New York: Nova Science.

Bjork, L., & Lindle, J. C. (2001). Superintendents and interest groups. *Educational Policy, 15*(1), 76–91.

Blackmore, J. (2009). Leadership for social justice: A transnational dialogue. *Journal of Research on Leadership Education, 4*(1), 1–10.

Brooks, J., & Miles, M. T. (2008). From scientific management to social justice . . . and back again? Pedagogical shifts in the study and practice of educational leadership. In A. H. Normore (Ed.), *Leadership for social justice: Promoting equity and excellence through inquiry and reflective practice* (pp. 94–114). Charlotte, NC: Information Age.

Brown, K. M. (2004). Leadership for social justice and equity: Weaving a transformative framework and pedagogy. *Educational Administration Quarterly, 40*(1), 77–108.

Callahan, R. E. (1966). *The superintendent of schools: A historical analysis*. East Lansing, MI: National Center for Research on Teacher Learning. (ERIC Document Reproduction Service No. ED 0104 410)

Coburn, C. E. (2006). Framing the problem of reading instruction: Using frame analysis to uncover the microprocesses of policy implementation. *American Educational Research Journal, 43*(3), 343–349.

Coburn, C. E., & Talbert, J. E. (2006). Conceptions of evidence use in school districts: Mapping the terrain. *American Journal of Education, 112*(4), 469–495.

Dantley, M. E., & Tillman, L. C. (2006). Social justice and moral transformative leadership. In C. Marshall & M. Oliva (Eds.), *Leadership for social justice: Making revolutions in education* (2nd ed., pp. 19–34). Boston, MA: Allyn & Bacon.

Darling-Hammond, L. (2007). Third annual Brown lecture in education research—The flat earth and education: How America's commitment to equity will determine our future. *Educational Researcher, 36*(6), 318–334.

DeMatthews, D. E. (2015). Making sense of social justice leadership: A case study of a principal's experiences to create a more inclusive school. *Leadership and Policy in Schools, 14*(2), 139–166.

DeMatthews, D. E. (2018). *Community engaged leadership for social justice: A critical approach in urban schools*. New York: Routledge.

DeMatthews, D. E., Izquierdo, E., & Knight, D. S. (2017). Righting past wrongs: A superintendent's social justice leadership for dual language education along the US-Mexico border. *Education Policy Analysis Archives, 25*, 1.

DeMatthews, D. E., & Mawhinney, H. (2014). Social justice leadership and inclusion: Exploring challenges in an urban district struggling to address inequities. *Educational Administration Quarterly, 50*(5), 844–881.

Evans, A. E. (2007). School leaders and their sensemaking about race and demographic change. *Educational Administration Quarterly, 43*(2), 159–188.

Furman, G. (2012). Social justice leadership as praxis developing capacities through preparation programs. *Educational Administration Quarterly, 48*(2), 191–229.

Glass, T. E., Bjork, L., & Brunner, C. C. (2000). *The study of the American superintendency, 2000: A look at the superintendent of education in the new millennium*. New York: R&L Education.

Green, T. L., & Dantley, M. E. (2013). The great white hope? Examining the white privilege and epistemology of an urban high school principal. *Journal of Cases in Educational Leadership, 16*(2), 82–92.

Green, T. L., & Gooden, M. A. (2014). Transforming out-of-school challenges into opportunities: Community schools reform in the urban Midwest. *Urban Education, 49*(8), 930–954.

Gregory, A., Skiba, R. J., & Noguera, P. A. (2010). The achievement gap and the discipline gap: Two sides of the same coin? *Educational Researcher, 39*(1), 59–68.

Ishimaru, A. (2013). From heroes to organizers: Principals and education organizing in urban school reform. *Educational Administration Quarterly, 49*(1), 3–51.

Kowalski, T. J. (2001). The future of local school governance: Implications for board members and superintendents. In C. Brunner & L. G. Björk (Eds.), *The new superintendency* (pp. 183–201). Oxford, UK: JAI, Elsevier Science.

Kowalski, T. J. (2005). *The school superintendent: Theory, practice, and cases.* Thousand Oaks, CA: Sage Publications.

Kowalski, T. J., McCord, R. S., Peterson, G. J., Young, P. I., & Ellerson, N. M. (2011). *The American school superintendent: 2010 decennial study.* New York: R&L Education.

Miles, M. B., & Huberman, A. M. (1994). *Qualitative data analysis: An expanded sourcebook.* Thousand Oaks, CA: Sage.

Park, V., Daly, A. J., & Guerra, A. W. (2013). Strategic framing: How leaders craft the meaning of data use for equity and learning. *Educational Policy, 27*(4), 645–675.

Rivera-McCutchen, R. L., & Watson, T. N. (2014). Leadership for social justice: It is a matter of trust. *Journal of Cases in Educational Leadership, 17*(4), 54–65.

Senge, P. M. (2006). *The fifth discipline: The art and practice of the learning organization.* New York: Double Day.

Spillane, J. P., Diamond, J. B., Burch, P., Hallett, T., Jita, L., & Zoltners, J. (2002). Managing in the middle: School leaders and the enactment of accountability policy. *Educational Policy, 16*(5), 731–762.

Spillane, J. P., Reiser, B. J., & Reimer, T. (2002). Policy implementation and cognition: Reframing and refocusing implementation research. *Review of Educational Research, 72*(3), 387–431.

Theoharis, G. (2007). Social justice educational leaders and resistance: Toward a theory of social justice leadership. *Educational Administration Quarterly, 43*(2), 221–258.

Weber, K., & Glynn, M. A. (2006). Making sense with institutions: Context, thought and action in Karl Weick's theory. *Organization Studies, 27*(11), 1639–1660.

Weick, K. E., Sutcliffe, K. M., & Obstfeld, D. (2005). Organizing and the process of sensemaking. *Organization Science, 16*(4), 409–421.

Woulfin, S. L., Donaldson, M. L., & Gonzales, R. (2016). District leaders' framing of educator evaluation policy. *Educational Administration Quarterly, 52*(1), 110–143.

# 8

# SYSTEM LEADERSHIP FOR CONTINUOUS IMPROVEMENT

## The Role of District-Level Leaders in Creating the Conditions for System-Wide Improvement

*Christina J. Dixon and David Eddy-Spicer*

In recent years, a small number of pioneering school districts have embraced continuous improvement as the fulcrum for improving student outcomes across their systems (David & Talbert, 2013; Park, Hironaka, Carver, & Nordstrum, 2013). Education researchers have likewise begun to recognize the promise of targeting the school district in fostering meaningful system-wide improvement in education (e.g., Daly & Finnigan, 2016; Honig, 2013). Both approaches restore the district or local educational authority to a primacy that has been embattled over the past two decades of market-oriented reforms in educational policy (Honig, 2008).

This renewed focus on system-wide improvement has been tightly linked to an emerging concern with system leadership (Hopkins et al., 2015), and the recognition of leaders' centrality in change efforts (Smylie, 2016). The attention to leadership of improvement features in the language of the Professional Standards for Educational Leaders (PSEL) (National Policy Board for Educational Administration, 2015). One of the standards highlights that practicing continuous improvement is an essential part of how educators must develop the field to "promote each student's academic success and well-being" (p. 18), and charges education leaders with the professional responsibility to aim improvement efforts squarely at success for all students. LeMahieu et al. (2017) also highlight the importance of leaders' efforts in enabling effective and sustainable improvement work to occur, explaining,

> [W]hen a problem occurs in the system (e.g., poor student outcomes), leadership has the role and responsibility to investigate systems-based causes. This involves trying to pinpoint the interactions among structures, work processes and norms that are producing the current outcomes.
>
> *(p. 15)*

The above characterization of system leadership contrasts starkly with traditional hierarchical bureaucracy and leadership through top-down mandates. Through the lens of continuous improvement, problems lie in largely tacit processes, persisting due to systemic causes. In a complex system like education, effectively addressing problems requires work within and across organizational levels. It further places primacy on process improvement and on capability development through collaborative problem solving, rather than hinging solely on leadership decisions to implement the right solution or an immediate fix. Leaders must facilitate and lead the critical examination of their systems in ways that encourage learning and continuous improvement to occur system-wide. Little definitive guidance exists, however, about how education leaders can effectively enact such systems thinking in practice.

This chapter seeks to contribute to the education field's understanding of how district-level leaders can best further district-wide continuous improvement efforts. We synthesize current research-based knowledge about leadership behaviors that create the conditions for successful continuous improvement. We are not concerned with bodies of research about effective education leadership practices in general or sources drawn primarily from school- or classroom-level contexts. Rather, our review emphasizes the role of superintendents and their leadership teams as the system-level leaders of school district improvement.

Available literature in education is limited. We draw on relevant findings regarding the effective leadership of quality improvement in multiple sectors along with an emerging research base around district-level improvement in education. Other industries have a more robust history of continuous improvement and have created bodies of guidance for leaders (e.g., Spear, 2009; Swensen, Pugh, McMullan, & Kabcenell, 2013). Our goal is to harvest relevant systems leadership guidance from other sectors and translate it for the leadership of continuous improvement in education.

## Continuous Improvement

The rapid growth of improvement efforts across multiple fields in recent years has led to wide variation in how the terms *quality improvement* and *continuous improvement* are currently understood and used across contexts. We use the definition articulated by Park et al. (2013) and subsequently elaborated by O'Day and Smith (2016) as described below. This definition offers practical criteria for distinguishing between systematic quality improvement approaches and other types of improvement due to its emphasis on understanding systems, using data strategically, and employing a specific improvement methodology.

Based on their comprehensive review of relevant literature and change efforts in multiple sectors, Park et al. (2013) identified five core characteristics of quality improvement, subsequently characterized by O'Day and Smith (2016) as:

1. focused on system outcomes for a defined population of beneficiaries –*and* on the processes that lead to these results;
2. using variation in performance (including "failure") as opportunities for learning and improvement;
3. taking a system perspective, with the understanding that systems are designed to get the results they produce, so if you want to change the results, you have to change the system;
4. evidence-based, including measurement of not only outcomes but processes (and resources); and
5. involving a specific and coherent methodology and processes (e.g., PDSA (Plan-Do-Study-Act) cycles, "Six Sigma," and "Lean").

(p. 315)

Continuous improvement, as defined here, consists of embedding quality improvement into the everyday work of individuals in a system (Park et al., 2013), with a focus on system-wide outcomes.

## Initial Framework and Methods

Our approach to rigorous review in this chapter is an adaptation of framework-based synthesis (Dixon-Woods, 2011).[1] We have selected a framework developed by the Institute for Healthcare Improvement (IHI), a respected leader of quality improvement in healthcare. Building on decades of research and practice, IHI's "Three Interdependent Dimensions of High-Impact Leadership in Health Care" (Swenson et al., 2013, p. 6) describes how leaders can best focus their efforts to achieve what IHI terms "Triple Aim" results: improve both the experience and outcomes of care while reducing cost. Swensen et al. (2013) offer three interdependent dimensions of leadership: the *mental models* that condition leaders' thinking around challenges and solutions; the *behaviors* that shape what leaders do to make a difference; and the places where leaders *focus* improvement efforts.

While the detailed content offered within this framework is particular to healthcare and may therefore not apply to education leaders, the overarching categories of leadership *mental models, behaviors*, and *focus* appeal for two reasons. First, they organize IHI's collective learning from experience and research as the organization has engaged healthcare leaders in systemic, ongoing change (Swensen et al., 2013). Second, IHI's thought leadership has been a primary source for scholars and practitioners seeking to practice quality improvement in education (Bryk et al., 2015). Consequently, IHI's thinking about leadership for improvement is likely to have utility for educators seeking to lead continuous improvement in their districts.

## Findings: Overview of Sources

Research concerning school districts intentionally engaging in continuous improvement is in its infancy, and primarily consists of individual or multiple case studies. Early contributions to the field include Elmore and Burney's (1998) thorough description of efforts in New York's District #2 that raised overall student performance and began to close the achievement gap in literacy, and McLaughlin and Talbert's (2003) four-year longitudinal study of how three California districts succeeded in closing the achievement gap faster than comparison districts. More recently, Park et al.'s (2013) study offers detailed, descriptive cases of school districts in Menomonee Falls, Wisconsin and Montgomery County, Maryland. Both districts made early attempts at adopting continuous improvement methods district-wide, and both have achieved measurable improvement in instruction and operations. Standing out for its thorough methods, David and Talbert's (2013) research describes Sanger Unified School District's approach to significantly outpacing California's rate of improvement on the Academic Performance Index and in increasing graduation rates among a student population that includes 73% low income and 84% minorities.

We also reviewed selected studies concerning the ways district-level leaders act to accelerate improvement, regardless of improvement approach. These sources are not specifically focused on leadership for continuous improvement in education; nonetheless their findings may have relevance for education leaders seeking to build a solid foundation for such work. The strongest of these studies focus deeply on multiple cases (Honig, 2013; Snipes, Doolittle, & Herlihy, 2002; Togneri & Anderson, 2003). In particular, Snipes, Doolittle and Herlihy's (2002) work stands out for its comparison of practices in improving and non-improving districts as opposed to relying on the more common outlier study design. Comprehensive literature reviews that include treatment of district leadership to improve student learning (Leithwood et al., 2004), the role of district central offices in improving instruction and student achievement (Mac Iver & Farley, 2003), and leadership for educational change (Schwahn & Spady, 2001) are also sources for this review.

The sources cited in healthcare and industry commonly draw on both research and practice to distill leadership lessons that deliver results. Notably, in addition to IHI's "High Impact Leadership" (Swensen et al., 2013, p. 6) framework that serves as the conceptual framework for this review, Steven Spear's *Chasing the Rabbit* (2009) mines multiple case studies of organizations across a wide variety of industries to distill commonalities that enable them to outpace their competitors in improving their ability to deliver results for their customers. Similarly, Toussaint and Ehrlich (2017) cite research on leadership for improvement as well as their learning from working with 40 healthcare CEOs to successfully improve system performance. This review also includes works published by organizations that distill leadership lessons from improvement literature and the practice of organizations that show remarkable results in their field, such as the Shingo Institute (Shingo, 2016).

In healthcare and other industries, the breadth and depth of research supporting our synthesis is sizeable and draws on decades of improvement work. Research in education, in contrast, is limited by the size of the research base and the design of available studies. Relatively few case studies are available, and of those only a very few examine more than one district. All focus on outlier districts that have achieved notable results, using a retrospective design to document progress. Furthermore, none focuses exclusively on the role of top district leaders and how their actions may create conditions for continuous improvement of student outcomes. For these reasons, the available research offers some suggestive evidence, but does not substantiate any cause and effect relationships between district conditions or leader behaviors and district-wide success.

The state of the research in the field of continuous improvement in education underscores the importance of examining related sources across sectors. Where findings from multiple studies and disciplines identify common effective leadership actions and characteristics, their strength increases with the number of cases that demonstrate a common result. Additionally, the duration and scale of the efforts outside of education make a strong case for the validity of identified leadership actions and characteristics in a given sector, and offer promising starting points for exploring the nature of leadership for improvement in education.

## Findings: Thematic Synthesis

The following thematic synthesis of findings from relevant sources across sectors is organized according to IHI's "High-Impact Leadership" (Swenson et al., 2013, p. 6) framework, and pinpoints where the literatures converge and conflict to describe the current state of knowledge regarding district-level leadership for continuous improvement in education. Findings are grouped by theme within each of IHI's three interdependent dimensions of leadership: "how leaders think about challenges and solutions" (*mental models*), "what leaders do to make a difference" (*behaviors*), and "where leaders need to focus efforts" (*focus*) (Swensen et al., 2013, p. 4).

### *How leaders think about challenges and solutions*

While much of the literature reviewed concentrated on leader actions, a significant subset enumerated mental models conducive to effective leadership of quality improvement. Literature from healthcare, industry and education all elaborated on the common themes that leaders must value learning, think systemically, respect every individual and embrace personal responsibility for their organization and its performance.

**Value learning.** From the early days of quality improvement's application in healthcare, the importance of learning has been emphasized for all, with a special role for leadership. In "A Primer on Leading the Improvement of Systems," IHI co-founder Don Berwick asserted, "the effective leader must understand that the road to improvement passes through change and that one efficient way to change is to learn from the actions we ourselves take" (1996, p. 312). Steven Spear (2009) echoes that view, concluding that leaders of continuously improving organizations must adopt the role of "Learner in Chief" (p. 294), modeling the learning mindset that must be cultivated in staff across the organization. Further explicating what it means to value learning, Toussaint and Ehrlich (2017) emphasize the importance for leaders to recognize that they do not have all the answers, to adopt a genuine willingness to change, and to model humility and curiosity in their interactions with others. These mindsets are prerequisites for leadership behaviors that support an organizational culture of improvement.

While not specifically focused on the leader's role, Lucas and Nacer's (2015) more recent review drawing primarily on psychological literature related to healthcare improvement identifies learning as one of "five core improvement habits" (p. 8). This research reinforces the importance of ensuring that the ability to develop this habit is embedded in daily work, which organizational leaders are uniquely positioned to do.

In education, Park et al. (2013) offer a rare multiple case study of continuously improving districts that supports the need for leaders to value their own learning. Among the cross-cutting themes identified in their analysis was that "leaders of continuous improvement bring a learning mindset to the work" (p. 23).

**Think systemically**. Playing a role in industry similar to that of IHI in healthcare, the Shingo Institute has synthesized the thought leadership of quality experts ranging from Ford, Juran, Deming and Ohno to Senge, Covey and Womack into a model for operational excellence (Shingo, 2016). Among the principles enumerated in this model is "think systemically," which the Shingo Institute defines as "understanding the relationships and interconnectedness within and between systems" (Shingo, 2016, p. 42).

Thinking systemically has also been recognized as important by improvers in the field of education (Shaked & Schechter, 2017). The authors of the PSEL standards included "adopt(ing) a systems perspective" as a key action of effective educators seeking to fulfill their school improvement standard (National Policy Board for Educational Administration, 2015, p. 18). Scholars at the Carnegie Foundation for the Advancement of Teaching have identified "seeing the system" as an essential principle of quality improvement (Bryk et al., 2015). Furthermore, they have noted that within continuous improvement organizations, problems of practice are understood as products of systems (Bryk et al., 2015) and that leaders have the responsibility to try to understand what causes their systems to produce their current results (LeMahieu et al., 2017). From a practical perspective, the case study of Montgomery County School District's continuous improvement efforts specifically calls out Superintendent Weast's insistence on "looking at the system as a whole" (Park et al., 2013, p. 15) as an important success factor. Furthermore, all three organizations studied by Park et al. insisted on adopting a "systems-thinking approach" (2013, p. 23) to their work that led them to recognize previously unnoticed interdependencies of key organizational processes. Other case studies of successful continuously improving districts similarly highlight the strategic importance of viewing the system as the unit of change, and ensuring that local improvements contribute to overall district health (Bisby et al., 2009; David & Talbert, 2013; Kirp, 2013; McLaughlin & Talbert, 2003).

**Respect every individual.** The Shingo Institute (2016) recognizes valuing every individual and supporting them in fulfilling their potential as another key principal of operational excellence, and counsels leaders to exercise humility and trust others to make good decisions. IHI also makes the assertion, based on an extensive review of the quality improvement literature and practitioner expertise, that leaders must embrace the idea that everyone in their organization is an improver (Swensen et al., 2013). Similarly, Toussaint and Ehrlich (2017) caution that leaders must embrace the reality that their organization's greatest asset is the person on the front line.

Schwahn and Spady (2001) go further to offer an articulation of the leader beliefs that support this value in their book, *Total Leaders*. This extensive review and synthesis of literature related to quality improvement and productive, sustained change in education finds that effective leaders believe that "a tremendous amount of power lies within each person and that their role is to create work environments that let that power and capability emerge" (Schwahn & Spady, 2001, p. 20). Similarly, Sanger Unified School District in California called out respect for people, starting with modeling by district-level leadership, as an essential element of their culture supporting continuous improvement (David & Talbert, 2013). Taken together, these sources point to the need for effective leaders of improvement to respect every person and value their contributions and potential to grow.

**Embrace personal responsibility.** The theme of leadership responsibility recurs throughout the literature (Berwick, 1996; Deming, 2000; LeMahieu et al., 2017; Schwahn & Spady, 2001; Shingo, 2016; Spear, 2009). Through the lens of continuous improvement, system leaders inhabit a role that offers a unique view across organizational siloes, along with the power to effect changes in the system as needed to accelerate progress. With that positioning, many argue, comes the responsibility to act intentionally to support the improvement efforts of people within the system. As Bryk asserted in his keynote delivered at the Carnegie Foundation's 2017 Summit on Improvement in Education, we should be "haunted" by the "predictable failures" of our systems" (Bryk, 2017, p. 10). Further supporting this view, Togneri and Anderson (2003) found that superintendents of rapidly improving districts accepted responsibility for their district's poor performance. Willingness to courageously acknowledge poor performance and question current practices appeared critical to superintendents' efforts to build will for change.

## *What leaders do to make a difference*

Across sectors, the literature examining approaches to leading quality improvement efforts is filled with recommended actions for leaders. This section of the review identifies the strongest, common themes regarding effective leader behaviors for improvement that emerge from both within and outside the field of education.

**Set a vision, purpose and strategy focused on results for students.** The literature regarding leadership for quality improvement across sectors consistently identifies setting and aligning the organization around a clear direction as an essential role of leaders (Berwick, 1996; Lunenberg, 2010; Park et al., 2013; Schwahn & Spady, 2001; Shingo, 2016; Snipes et al., 2002; Spear, 2009; Swensen et al., 2013; Togneri & Anderson, 2003). Furthermore, it is the most consistent finding across the education literature regarding what leaders do to support continuous improvement (Bisby et al., 2009; David & Talbert, 2013; Kirp, 2013; McLaughlin & Talbert, 2003; Park et al., 2013). In keeping with quality improvement leader W. Edwards Deming's counsel to "create constancy of purpose" (2000, p. 23), leaders are instructed to establish a clear and compelling aim and purpose (Berwick, 1996; Schwahn & Spady, 2001; Shingo, 2016), maintain "relentless focus" (Swensen et al., 2013, p. 86) and build will around a shared vision (Schwahn & Spady, 2001; Swensen et al., 2013). This includes aligning the organizational design and providing the conditions to support the achievement of that vision (Spear, 2009; Swensen et al., 2013).

The education literature further specifies the essential elements of a vision to drive continuous improvement in schools and districts. As previously mentioned, the Professional Standards for Education Leaders (National Policy Board for Educational Administration, 2015) stipulate that the goal of continuous improvement is to "promote *each* student's academic success and well-being" (p. 18), arguing that in education, the compelling purpose worthy of relentless focus across the field is to drive better outcomes for every child. Aligned with this standard, all of the case studies of continuously improving districts identified the need for the superintendent to shepherd the establishment of a common vision and strategy that includes a long-term commitment to continuous improvement focused on improving student learning (Bisby et al., 2009; David & Talbert, 2013; Kirp, 2013; McLaughlin & Talbert, 2003; Park et al., 2013). According to many of these studies, goals must be unambiguous, and district leaders have the responsibility to ensure that they carefully choose a small number of well-integrated priorities and strategies that form a cohesive system for achieving those goals (Bisby et al., 2009; David & Talbert, 2013; Kirp, 2013).

**Develop capability.** The leadership behavior that figures most prominently is to develop the core capabilities of improvement and the skills to lead the development of improvement capabilities in others (e.g., David & Talbert, 2013; Honig, 2013; Leithwood, 2010; McLaughlin & Talbert, 2003, National Policy Board, 2015; Park et al., 2013; Spear, 2009; Swensen et al., 2013). Improvement leaders have the responsibility to both directly support the growth and learning of their people, and to establish "the cascade of capability development throughout the organization" (Shingo, 2016, p. 38). Spear (2009) describes how developing individual capability across an organization yields, "organizational capacity to be self-correcting, self-improving, and self-innovating" (p. 264).

Those promoting and engaged in continuous improvement in the field of education confirm this emphasis. PSEL standards (National Policy Board for Educational Administration, 2015) describe effective leaders as those who develop skills and capability, and promote their staff's leadership of improvement. The continuously improving district case studies similarly emphasize the need for district leaders to focus heavily on building capability for themselves and their school level staff, or as David and Talbert (2013) describe it, to "shift to leadership for learning" (p. 23). They must learn to grow principals as instructional leaders and improvers, and in turn, principals must learn to grow the capabilities of teachers (David & Talbert, 2013).

In the education literature, the focus on developing individual capability in service of building organizational capacity extends beyond the superintendent to the central office. According to multiple case studies, central offices need to engage in improving their own work to become capable of modeling the inquiry and risk-taking behavior needed for the rapid learning that continuous improvement requires (Agullard & Goughnour 2006; David & Talbert, 2013; Kirp, 2013; McLaughlin & Talbert, 2003; Park et al., 2013). Perhaps most strikingly, Honig's 2013 study of effective central office practices for helping schools to dramatically improve student learning outcomes identified the need for "superintendent and other central office leadership that will help staff continuously build their capacity for better performance" (p. 1). In her extensive observations of three school systems engaged in such transformation, Honig recognized patterns in leader behavior that included hands-on coaching and teaching of staff toward continuously improving their work to meet new system demands. The experiences of the districts studied suggest that leaders must not only provide for, but actively lead the development of their people for continuous improvement to flourish.

**Create a culture of improvement**. The notion that leaders have primary responsibility for creating an organizational culture that supports improvement figures prominently in improvement literature from healthcare and industry (Kaplan, Provost, Froehle, & Margolis, 2012; Shingo, 2016; Swensen et al., 2013), as well as in the majority of case studies of continuously improving districts (David & Talbert, 2013; Elmore & Burney, 1998; Kirp, 2013). Shingo's (2016) framework identifies driving and managing culture as a lever for top leaders in driving improvement-aligned behaviors in their staff. Likewise, the Model for Understanding Success in Quality (MUSIQ) in healthcare (Kaplan et al., 2012) calls out creating a culture (values, norms and beliefs) that supports staff in pursuing quality improvement as a key factor in successful improvement efforts. Guidance for how leaders can act to create such a culture includes noticing and calling attention to specific behaviors they observe that are close to the ideal (Shingo, 2016), communicating and modeling desired behaviors (Swensen et al., 2013), ensuring transparency (Swensen et al., 2013), practicing reflection (Berwick, 1996), asking effective, open-ended questions (Toussaint & Ehrlich, 2017), engaging in active listening (Swensen et al., 2013), and showing a willingness to change upon learning something new (Shingo, 2016).

In the field of education, the literature concerning continuously improving school districts repeatedly identifies the need for the central office to nurture a district-wide culture supportive of such improvement (David & Talbert, 2013; Elmore & Burney, 1998; Kirp, 2013), although the features of this culture deemed most important vary significantly across districts. Sanger emphasizes positive relationships, respect for people, collaboration and shared responsibility, and "reciprocal accountability" that honors professionalism and is supportive rather than punitive (David & Talbert, 2013, p. 27). Others note the importance of cultivating trust (Agullard & Goughnour, 2006; Kirp, 2013; McLaughlin & Talbert, 2003), the need for transparency (Elmore & Burney, 1998) and open communication (Agullard & Goughnour, 2006).

**Span boundaries.** The leadership function referred to as "boundary spanning responsibility" (Lunenberg, 2010; Spear, 2009) is defined by Spear as being "system-oriented – responsible for the design and operation of processes at levels of aggregation for which others have insufficient perspective and authority" (p. 263). Swensen et al. (2013) call this responsibility "boundarilessness" (p. 4) while recommending that leaders "model and encourage team work and systems thinking" (p. 5).

The literature highlights two ways that leaders can practice boundary spanning behavior. The first is to set a vision for changing the system, rather than merely making change within it (Berwick, 1996; Honig, 2013; Lunenberg, 2010). Berwick (1996) asserts that leaders must challenge the status quo and advocate for a clear alternative to the current system. In practice, Honig's (2013) observations of successfully transforming central offices led her to conclude that effective leadership "sets a vision for ambitious, performance oriented change that moves beyond tinkering to true transformation" (p. 9). Second, leaders must both break down siloes and build others' capability to work across them to solve problems (Honig, 2013; Park et al., 2013). Due to districts' organizational complexity, many entrenched problems lie at the intersection of traditional departments and organizational levels. Therefore, all staff must learn to collaboratively redesign processes to achieve desired outcomes rather than conform to custom. Leaders must model and drive this learning to enable continuous improvement to take hold.

## Where Leaders Need to Focus Efforts

In regard to the question of where leaders can best direct their attention and energy in accelerating improvement, the literature suggests the importance of leaders creating organizational conditions supportive of continuous improvement. Within this area of focus, promoting organizational alignment and creating an effective improvement infrastructure are two key areas needing targeted leader attention.

**Promote organizational alignment.** Perhaps the simplest description of this focal issue comes from the Shingo Institute, which asserts that leaders ensure "that the systems are designed so that it is easier to do the right thing than the wrong thing" (2016, p. 40). This statement implies a recognition that often work systems are not built to encourage desired behavior, and that it is the responsibility of leadership to fix this problem where it exists.

The education literature offers significant evidence of the importance of aligning district infrastructure and ensuring coherence across all elements of the instructional core for successful continuous improvement (e.g., O'Day & Smith, 2016; Snipes et al., 2002; Togneri & Anderson, 2003). Honig's (2013) study of three successfully transforming districts offers an example of how this can be done, suggesting "transformation should involve. . . developing and aligning performance-oriented central office services to support district-wide instructional improvement" (p. 1). As these and other continuously improving districts have learned, district offices must do the hard, ongoing work needed to engage staff across the district in aligning all structures, programs and processes with the district vision and goals, and to integrate new initiatives into the existing strategy in ways that maintain its coherence (Bisby et al., 2009; David & Talbert, 2013; Honig, 2013; Kirp, 2013; McLaughlin & Talbert, 2003; O'Day & Smith, 2016; Park et al., 2013). As Childress,

Elmore & Grossman (2006) observed based on their intensive engagement with 15 improving districts across the United States, "success depends on the degree to which the strategy is implemented consistently across an entire system over several years. Doing this requires the creation of an organization whose culture, systems and structures, resources, and stakeholder relationships support the strategy" (Creating a Coherent Organization, para. 12).

One notable aspect of alignment particular to school districts concerns the role of school boards. Although school boards are absent from the many case studies of continuously improving districts, those that do mention them make a case for the importance of solid board support to the success of a district's continuous improvement efforts. Superintendents in both Montgomery and Menomonee Falls stressed the importance of aligning with the board around a common vision and goals (Park et al., 2013), and Sanger's district leaders described purposeful efforts to nurture strong relationships with school board members and maintain their solid, sustained support for continuous improvement (David & Talbert, 2013). Reinforcing these leader perceptions, McLaughlin and Talbert's (2003) evidence from surveys and interviews showed that district administrator ratings of school board support for the reform efforts were the only context variable studied that proved to be a significant predictor of whether the district succeeded in developing the central office as a learning organization and providing instructional support to schools.

**Create an effective improvement infrastructure.** The literature reviewed offers multiple examples of how leaders can build organizational systems and processes to support improvement. High level recommendations include "integrate improvement work into daily work at all levels" (Swensen et al., 2013, p. 23) and creating and using feedback loops to improve performance (Schwahn & Spady, 2001). More concrete recommendations common in the literature involve establishing disciplined improvement methods, supporting the effective use of data, and developing leaders from within the district.

***Disciplined improvement methods.*** While their specific methodologies differed, the district leaders studied established disciplined improvement processes to embed quality principles into the way work was done system-wide. These districts developed their staff's capability to use a shared improvement approach, whether they conducted improvement cycles (Elmore & Burney, 1998; Park et al., 2013) or Cycles of Inquiry (McLaughlin & Talbert, 2003), formed improvement-focused Professional Learning Communities (David & Talbert, 2013), or adopted the Plan-Do-Check-Act (PDCA) cycle, Six Sigma or Lean (Bisby et al., 2009; Park et al., 2013).

***Effective use of data.*** Key to this approach is that "decisions must be grounded in evidence" (David & Talbert, 2013, p. 8), a requirement that needs to be supported by a robust and aligned, district-level measurement and data analysis infrastructure (Agullard & Goughnour 2006; Kirp, 2013; McLaughlin & Talbert, 2003; Park et al., 2013). Leaders need to ensure that appropriate data systems (National Policy Board, 2015) and professional development prepare and support district personnel at all levels in making data-driven decisions (Mac Iver & Farley, 2003; Snipes et al., 2002; Togneri & Anderson, 2003). Specific recommended uses of data for improvement include using data to monitor progress toward goals (Park et al., 2013; Waters & Marzano, 2006); planning, organizational learning and accountability (Leithwood, 2010); and shifting the focus of evaluation to promote improvement (Park et al., 2013). This last point may represent the greatest departure from the status quo of data use in most school districts. It requires a change in focus from accountability to learning, and an emphasis on formative and process-level data as an essential complement to the more common use of summative outcomes data.

***Developing leaders from within.*** One other key role of district leaders prominent in the experience of a couple of improving districts was the need for them to create intentional pipelines to grow new leaders from within. Recognizing that having the experience of doing continuous improvement is important to learning how to lead it effectively, Sanger integrated a focus on leadership development into its strategy from the beginning (David & Talbert, 2013). New York's District #2 also had an intentional plan to develop leaders internally, and when it failed to produce improvement savvy leaders at the rate needed by the district, Elmore and Burney (1998) called out that shortcoming as a specific threat to the long-term sustainability of the district's continuous improvement efforts.

## A Revised Framework

The thematic synthesis enables us to elaborate our initial framework. The revised framework shown in Figure 8.1 below should be read from bottom to top, with each layer building on the foundation laid by the one preceding it.

The above expanded framework represents an initial model of how education leaders initiate and sustain continuous improvement across their districts and offers a starting place for further inquiry. According to this hypothesis, leader mental models set a foundation upon which system-wide continuous improvement can be built, motivating and enabling leaders to act in ways that are conducive to the effective leadership of continuous improvement efforts. In turn, specific leader behaviors, when aligned with the right mental models, can create fertile ground for improvement to take root. Further, the most successful district-level leaders focus their efforts on high leverage organizational challenges that they are uniquely positioned to address. Together, these mental models, behaviors and areas of focus enable such leaders to create the conditions for continuous improvement to become embedded in everyone's work, every day, district-wide. Doing work in this way then produces sustained progress in improving student outcomes and the system that produces them.

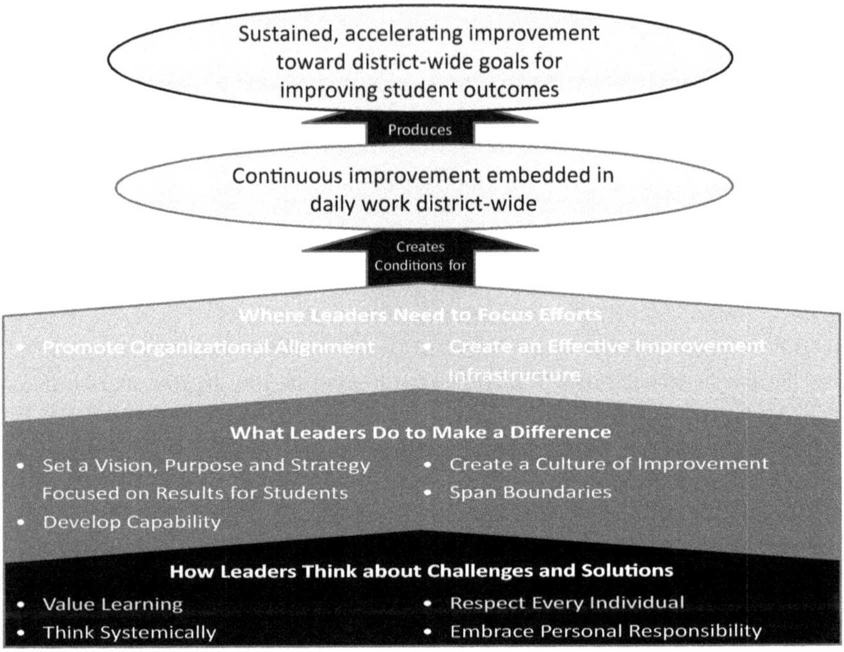


**FIGURE 8.1** Revised Framework for District-level Leadership of Continuous Improvement.

Adapted from *Executive Leadership for Continuous Improvement in K-12 Public School Districts: A Comparative Case Study* (Unpublished Capstone Proposal), by Christina J. Dixon, 2017, University of Virginia, Charlottesville, VA.

## Conclusion

The literatures within and outside of education generally converge around common themes related to the nature of leadership for continuous improvement. This convergence suggests that literature from outside of education may bear important messages for our understanding of system leadership in education. That said, the evidence in education is very limited, and additional field research is needed to focus specifically on understanding the practice of leadership of continuous improvement district-wide. This synthesis of the literature highlights several compelling questions for further study, including: What are the mental models and moves of leaders who successfully initiate district-wide continuous improvement? How do effective leaders prioritize their actions? What do they do to overcome common barriers?

Continuous improvement and systems thinking have both been identified as promising approaches for educators seeking to address complex, systemic challenges and sustainably improve student outcomes (O'Day & Smith, 2016; Shaked & Schechter, 2017). The findings of this inquiry suggest that systems thinking is an essential component of district-level leadership of continuous improvement, and that continuous improvement may provide an approach for leaders seeking to enact systems thinking and reap rewards district-wide. Therefore, there is likely benefit in investigating the nature of the intersection between systems thinking and continuous improvement and its practical implications for the field. How can efforts to practice systems thinking and continuous improvement at the district level inform each other? In what ways could such efforts be integrated to accelerate the achievement of student outcomes system-wide?

The contributions of this work need not be limited to the field of education, but may add to the fields of systems thinking and continuous improvement more broadly. Future inquiry could also include investigating the questions of whether and how the special attributes of education systems could enable the field of education to expand or elaborate on what is known about leadership of continuous improvement within complex systems. We have much to gain – and much to learn – from the rich theoretical, empirical, and practical wisdom offered through the enactment of systems thinking in the leadership of system-wide continuous improvement in education.

## Note

1 Framework-based synthesis is a form of theory-driven review, which Gough, Sandy and Thomas (2012) characterize as "configurative" approaches that are most useful for elaborating and testing patterns drawn from a widely varied body of literature. Framework synthesis begins with the identification of an initial theory that enables the reviewer to identify themes a priori, and then through the process of review, to test the initial framework against the literature.

## References

Agullard, K. & Goughnour, D. (2006). *Central office inquiry: Assessing organization, roles, and functions to support school improvement.* San Francisco: WestEd.

Berwick, D. M. (1996). A primer on leading the improvement of systems. *British Medical Journal, 312*(7031), 619–622.

Bisby, D., Davis, C., Hall, W., Muzzy, S., & Schafer, K. (2009). *Achieving and sustaining excellence in education.* Milwaukee, WI: American Society for Quality. Retrieved from http://207.67.83.164/edu/2009/07/continuous-improvement/achieving-and-sustaining-excellence-in-education.pdf

Bryk, A. S. (2017). Speech presented at Fourth Annual Carnegie Foundation Summit on Improvement in Education: *Redressing Inequities: An Aspiration in Search of a Method.* San Francisco, CA: Carnegie Foundation for the Advancement of Teaching.

Bryk, A., Gomez, L., Grunow, A., & LeMahieu, P. (2015). *Learning to improve: How America's schools can get better at getting better.* Cambridge, MA: Harvard Education Press.

Childress, S., Elmore, R., & Grossman, A. S. (2006). How to manage urban school districts. *Harvard Business Review, 84*(11), 55–68. Retrieved from https://hbr.org/2006/11/how-to-manage-urban-school-districts

Daly, A. J., & Finnigan, K. S. (Eds.) (2016). *Thinking and acting systemically: Improving school districts under pressure*. Washington, DC: American Educational Research Association.

David, J., & Talbert, J. (2013). *Turning around a high poverty district: Learning from Sanger*. San Francisco, CA: SH Cowell Foundation.

Deming, W. E. (2000). *Out of the crisis* (3rd ed.). Cambridge, MA: MIT Press.

Dixon, C. J. (2017). *Executive leadership for continuous improvement in k-12 public school districts: A comparative case study*. (Unpublished capstone proposal). University of Virginia, Charlottesville, VA.

Dixon-Woods, M. (2011). Using framework-based synthesis for conducting reviews of qualitative studies. *BMC medicine, 9*, 39–39. doi:10.1186/1741-7015-9-39

Elmore, R., & Burney, D. (1998). *Continuous improvement in community school district # 2, New York City*. Pittsburgh, PA: University of Pittsburgh, Learning Research and Development Center.

Gough, D., Sandy, O., & Thomas, J. (2012). *An introduction to systematic reviews*. London; Thousand Oaks, CA: Sage.

Honig, M. I. (2008). District central offices as learning organizations: How sociocultural and organizational learning theories elaborate district central office administrators' participation in teaching and learning improvement efforts. *American Journal of Education, 114*, 627–664.

Honig, M. I. (2013). From tinkering to transformation: Strengthening school district central office performance. *Education Outlook, 4*, 1–10. Washington DC: American Enterprise Institute for Public Policy Research.

Hopkins, D., Stringfield, S., Harris, A., Stoll, L. & MacKay, T. (2015). School and system improvement: A narrative state-of-the-art review. *School Effectiveness and School Improvement: An International Journal of Research, Policy and Practice, 25*(2), 257–281.

Kaplan, H. C., Provost, L. P., Froehle, C. M., & Margolis, P. A. (2012). The model for understanding success in quality (MUSIQ): Building a theory of context in healthcare quality improvement. *BMJ Quality & Safety, 21*, 13–21. Doi:10.1136/bmjqs-2001-000010

Kirp, D. L. (2013). *Improbably scholars*. Oxford, UK: Oxford University Press.

Leithwood, K. (2010). Characteristics of school districts that are exceptionally effective at closing the achievement gap. *Leadership and Policy in Schools, 9*, 245–291.

Leithwood, K., Louis, K., Anderson, S., & Wahlstrom, K. (2004). *How leadership influences student learning: A review of the evidence linking leadership to student learning*. New York, NY: The Wallace Foundation.

LeMahieu, P. G., Grunow, A., Baker, L., Nordstrum, L. E., & Gomez, L. M. (2017). Working to improve: Seven approaches to quality improvement in education. *Quality Assurance in Education, 25*(1), 5–25.

Lucas, B., & Nacer, H. (2015). *The habits of an improver*. London, UK: The Health Foundation.

Lunenberg, F. (2010). Total quality management applied to schools. *Schooling, 1*(1), 1–6.

Mac Iver, M. A., & Farley, E. (2003). *Bringing the district back in: The role of the central office in improving instruction and student achievement*. Baltimore, MD: Center for Research on the Education of Students Places at Risk.

McLaughlin, M., & Talbert, J. (2003). *Reforming districts: How districts support school reform*. Seattle, WA: Center for the Study of Teaching and Policy.

National Policy Board for Educational Administration (2015). *Professional Standards for Educational Leaders 2015*. Reston, VA: Author.

O'Day, J. A., & Smith, M. S. (2016). Quality and equality in American education: Systemic problems, systemic solutions. In I. Kirsch, H. Braun (Eds.), *The dynamics of opportunity in America* (pp. 297–358). Princeton, NJ: Educational Testing Service.

Park, S., Hironaka, S., Carver, P., & Nordstrum, L. (2013). *Continuous improvement in education.* Stanford, CA: The Carnegie Foundation for the Advancement of Teaching.

Schwahn, C. J., & Spady, W. (2001). *Total leaders: Applying the best future-focused change strategies to education.* Lanham, MD: The Scarecrow Press, Inc.

Shaked, H., & Schechter, C. (2017). *Systems thinking for school leaders: Leadership for excellence in education.* Cham, Switzerland: Springer International Publishing.

Shingo Institute (2016). *The discover excellence participant guide* (version 2.9). Logan, UT: Utah State University.

Smylie, M. (2016). Commentary: Three organizational lessons for school district improvement. In K. S. Finnigan & A. J. Daly (Eds.) *Thinking and acting systemically: Improving school districts under pressure* (pp. 209–220). Washington, DC: American Educational Research Association.

Snipes, J., Doolittle, F., & Herlihy, C. (2002). *Foundations for success: Case studies of how urban school systems improve student achievement.* Washington, DC: Council of the Great City Schools.

Spear, S. J. (2009). *Chasing the rabbit: How market leaders outdistance the competition and how great companies can catch up and win.* New York: McGraw-Hill.

Swensen, S., Pugh, M., McMullan, C., & Kabcenell, A. (2013). *High impact leadership: improve care, improve the health of populations, and reduce costs.* Cambridge, MA: Institute for Healthcare Improvement.

Togneri, W., & Anderson, S. E. (2003). *Beyond islands of excellence: What districts can do to improve instruction and achievement in all schools.* Washington, DC: Learning First Alliance.

Toussaint, J. S., & Ehrlich, S. P. (2017). Five changes great leaders make to develop an improvement culture. *NEJM Catalyst.* Retrieved from https://catalyst.nejm.org/five-changes-great-leaders-improvement-culture/

Waters, T. J., & Marzano, R. J. (2006). *School district leadership that works: The effect of superintendent leadership on student achievement. A working paper.* Denver, CO: Mid-Continent Research for Education and Learning (McREL).

# PART III

# Systems Thinking at the State/National Level

# 9

# THE IMPACT OF THE "SOCIAL" IN SOCIAL MEDIA SPACE

## A Systems Perspective on Educational Policy and Leadership

*Alan J. Daly, Miguel del Fresno, and Jonathan A. Supovitz*

### Interconnected Systems

We are social, meaning-making creatures and have been since the dawn of time. In fact, our survival and evolution were based squarely on the idea that we looked out for one another and worked together to shelter ourselves, hunt for food, and raise families. While we did so in a decidedly off-line world, our lives today are no less social – it is just that in the contemporary day we exist in a wider social system that spans both on- and offline worlds. In fact, research suggests that we are part of a much larger interconnected social system with emergent structures, ongoing interactions between the individual and collective, and the formation of patterns within the seeming chaos. Although we can think about systems and systems leadership in terms of classrooms, schools, and districts, in this modern age the idea of a systems approach must consider the wider virtual world in which we live and work. Let us consider some statistics on the scope of the system about which we are discussing.

IBM estimates that we create 2,500,000,000,000,000,000 (2.5 quintillion bytes) of information each day (see http://www-01.ibm.com/software/data/bigdata/what-is-big-data.html for an example). A number that size is difficult to imagine, so let us put that number in perspective. The data created every day, that 2.5 quintillion bytes, would fill 10 million Blu Ray Discs, which if stacked on top of one another would be as tall as 4 Eiffel Towers. The *volume* of that amount of data is staggering, but not only is the volume impressive, the *velocity* at which it is created is equally staggering. Data *velocity* estimates suggest that for every minute of every day there are 204,000,000 emails sent, 72 hours of YouTube video uploaded, 216,000 Instagram photos posted, and most importantly for our project, around 300,000 tweets tweeted. The volume and velocity of the data is incredible, but the *variety* of the data is equally impressive.

Within any 24-hour period the data generated can include: text, audio, video, click streams, sensors, and a host of other forms that get entered by human, machine or bot. Out of all that production, IBM estimates that 90% of the data is "unstructured," meaning it is a seemingly random collection of photos, tweets, and logs that are not ordered in any particular manner, nor organized for easy analysis and often ends up looking like a chaotic jumble – which makes the job of understanding this set of interactions as a system very challenging. These facts indicate that we live in an increasingly connected and interconnected system, and given this reality, we need new and unique ways to start to make sense of it all and search for important signals in the noise.

The Internet and its architecture, inherited from its initial design from the beginning of the 1990s, has allowed the evolution to Web 2.0, as well as the development and massive use of social platforms, known as *social media*. Social media are online technological platforms created to facilitate different ways of personal interaction – synchronous and asynchronous – with a global and local reach. Social media allows people to connect, share and discuss any topic, in any place and at any time, creating a vast communication system that is catalyzed through the ubiquity of devices.

The objective of this chapter is to provide an example of the much larger system that is at play around the educational endeavor, whether that be in the policy, practice, or leadership realm. This chapter draws on our much larger study that was published online (Supovitz, Daly, & del Fresno, 2015; see hashtagcommoncore.com) and only reflects a small part of the overall project which covered a 24-month period. We have selected this data set to share in an effort to illustrate the idea that there is a much larger social structural system that surrounds educational policy making processes and discourse and to heighten the awareness of policy makers, educators, and leaders of the larger system in which their work is embedded. This research is the first of its kind to use a unique blend of technology and methods as applied to large-scale data around an educational policy. In this chapter we are wanting to represent this larger system in a graphic manner to catalyze a larger conversation around the idea of how we conceptualize and examine the idea of "system."

Our work is grounded in *Social Network Analysis* (SNA), which enables us to examine systems at levels that go well beyond what we were able to do in the recent past. Given this new ability we now have the opportunity to expand how we think about systems and how policymakers and leaders may understand, act and interact within these systems. We now can capture, analyze, and represent network interactions on a scale that was not available before and open up new ways of conceptualizing systems. In this chapter we will focus on one of the most popular and much discussed social media platforms, Twitter, given its ubiquity and interest in the educational space. We will examine the US policy around the *Common Core State Standards (CCSS)* as it is one of the most ambitious and controversial educational policies in the United States as well as having day to day impact on educators and educational leaders.

## Social Media Mining, Social Network Analysis, and Twitter as a Vehicle

Social media exemplifies the extraordinary capacity of people to generate, spread, and exchange ideas, information, and opinions through interpersonal collective communication within a massive socially networked system. Social media can be organized into multiple different general categories: platforms of social *networking* (Facebook, LinkedIn), *microblogging* (Twitter, Weibo), *photography* (Flickr, Instagram, Pinterest), *video* (YouTube, Vimeo, MetaCafe), *social news* (Meneame, Digg, Reddit), *direct streaming* (Livecast, Ustream), *social gaming* (World of Craft), *bookmarking* (Delicious, StumbleUpon), *blogs* (Wordpress, Blogger), etc. Social media has become a source of large-scale 'big data' that is constantly being produced and offers opportunities and challenges to researchers. It is not so much the "bigness" of the data, but the fact that it reflects a larger system we can mine and examine for larger patterns that go beyond discreet parts. In order to capture this ever growing data we must be able to "mine" it in ways that allow for analysis.

Social Media Mining (SMM) can be described as the process of extracting, storing, representing, visualizing and analyzing data generated by users with the objective of discovering significant patterns (the spread of information, influence, social or consumption behavior, etc.) from the social interactions in online social media (Guandong & Lin, 2013; Jinhui, Meng, Xian-Sheng, & Tat-Seng, 2012; Zafarani, Abbasi, & Liu, 2014). We mine the data to deepen the understanding, both real and potential, of patterns of interaction, and to identify the structure and individuals with disproportionate influence (often called *superhubs*) within the larger social system.

Social media data mining increases the challenges of collection, analysis, and representation of data as these sources for a variety of reasons: 1) The data in social media is generated by the activity of the people in their online social platforms through multiple types of interactions; 2) Data continues to grow, multiply and disappear each day, because the sources are changing which means a constant technical and methodological challenge to following a flow; and 3) The diversity of the sources increases the difficulty of data aggregation in a consistent manner that allows the search and identification of significant patterns. Each of these areas presents challenges to tackling the system and makes this work difficult, but it is also the space in which some of the greatest research opportunities exist for thinking even more broadly about what we mean by "systems" and how we define and make sense of these systems.

## The Role of Activity and Social Space

At its heart, the work presented in this chapter draws on social media and is based on the proposition that activity in social space produces complex systems that are situated in networks of interactions and interdependence. In our work we privilege social interactions and relationships as the starting point of the analysis. This represents a departure from how many research endeavors are undertaken. Typically, policy studies draw on attributes of individuals, their formal positions or even some outcome measure, and then undertake a line of inquiry and analysis of which relations may be a part. In this work we flipped that idea and started from the constellation of interactions, thought of as a kind of social behavioral activity, and then moved out into concepts such as individual attributes and formal roles. In this manner, we let the social behavior of the actors drive our analysis as opposed to what we anticipated the activity might be based on individual level attributes. To be clear, we are not discounting individual attributes and affiliations, rather we see the social and individual perspectives to be complementary, each bringing something to our understanding of the larger system.

In this work we are interested in social "systems" that start at the dyadic (pair), faction (group), and system (entire network) level. This multi-level system approach recognizes the larger social milieu in which individuals communicate and interact and represents a different starting point suggesting the shift from dedicated work on lone individuals/institutions and their attributes to exploring the larger *systems* in which individuals engage around policy, practice, and leadership.

The science behind this approach is both solid and evolving. The work is grounded in social network theory and analysis (Scott, 2000). Contemporary social network theory and analysis have been argued to have been formalized in the 1930s with the work of Moreno (1934), who termed the field "sociometry." The key idea from this work is that one's position in a social structure has consequences for that individual, and the larger social system, which arises from a set of interactions, is consequential for an entire system. The balance of network research has demonstrated in slightly different ways that the structure of the larger network and an actor's position within that network are consequential to the overall network as well as the individual. It is this structure of a social system generated through social network analysis that is the core focus of this chapter.

Social network analysis (SNA) provides insight into how the social processes involved in communication and diffusion of ideas/opinions are stretched across individuals and levels of a system (Daly, 2010). Social network studies primarily focus on how the constellation of relationships in networks may facilitate and constrain the flow of "relational resources" (attitudes, beliefs, knowledge, materials, etc.), as well as providing insight into how individuals gain access to, are influenced by, and leverage these resources. The network perspective does not supplant the importance of individual attributes in understanding the selection, interpretation, and use of data, but rather offers a complimentary perspective and set of methods for better understanding the dynamic influence of social processes.

Rather than trying to understand the world based on the attributes of an actor (gender, years of experience, training, education, beliefs, formal position etc.), we focused on the influence and outcome of an individual's position vis-à-vis social ties with others, as well as the overall social structure of a network. In many cases, social network theorists suggest that the underlying social structure determines the type, access, and flow of resources to actors in the network, leading some scholars to suggest that the old adage "It is not what you know, but who you know," would more accurately be "Who you know defines what you know" (Cross & Parker, 2004).

The idea of interdependency is central to the notion of systems thinking in this regard and is critical to the way we approach our work from a more social perspective. Social embeddedness, in a network sense, refers to the nested nature of relations in a network (Granovetter, 1983, 1985). In a social network, individuals are embedded within pairs of relationships, and these pairs are embedded in larger groups of three, four, or more individuals that eventually form a social network. Even a social network itself is embedded in a larger social structure, for instance a community or a country. Social embeddedness also implies that changes at a single level (e.g., the pair) will have consequences for a higher-order level (e.g., the whole network) and in turn the larger structure of the network influences an individual's ability to access resources. As such, the significance of the pair extends beyond just those two individuals into a larger eco-system of connections. It is through a better understanding of this larger interconnected and interdependent network that we argue large-scale policy can be better understood and engaged.

The concept driving the work in this chapter is a more relational approach to studying systems surrounding a large-scale policy. Moving this more interactional perspective forward means that we need to move beyond vestiges of an industrial, mechanistic age in which people were viewed as only playing a formal and predictable role to viewing the world as a larger, interconnected and dynamic system. Conceptualized broadly, the idea is an intuitive one. Relationships matter in a very central manner to the ways in which information, knowledge, and opinion are moved. Therefore, the work of understanding policy in virtual space is at its core social and systems work.

## Twitter as a Vehicle

Twitter (www.twitter.com) is a global and free online social media platform where users write short messages with limited characters, known as *tweets*, which are sent to all of whom have opted to receive the tweets of the writer (*followers*). When posting a tweet (via web, text, computer, tablet or a smart phone) it is possible to add more than text, you can also include images, links and hashtags (a word or phrase prefixed by the symbol # that turns them into metadata). Currently Twitter is the fastest, simplest and cheapest social media platform where there is a flow of all types of information, news, ideas, events, rumors, multimedia, etc. created by users and moved quickly. This combines to make Twitter a privileged field for research, as it is a type of "central nervous system" of the Internet (Dorsey, 2012), as well as lending itself to research related to social structure (del Fresno, 2014).

From SNA's point of view the most significant aspect of Twitter for research is that it facilitates: 1) the communication and exchange of massive amounts of data and information simultaneously, and 2) at least two types of explicit relationships exist and can be differentiated: *retweets* (RTs) and *mentions or answers* (MTs), which we will focus on for this chapter. These two types of explicit relationships are more significant than *followers* – *following* relationships, this because RTs and MTs are a junction of explicit connections within the same network. RTs are the number of times a tweet of an individual "RT @username" is forwarded to the *timeline* of all of the followers of that user. Receiving a large number of RTs is argued to reflect the ability of an individual in his/her role as producer (micro medium) of significant information for others and can be used as a measurement of influence. We will use a measure of outdegree, outgoing ties, to analyze the interactions around RTs. MTs (mentions and answers) are the number of times that an individual shows up in the tweets of others as "@username." A large number of MTs indicates an established dialogue with others and reflects another measurement of influence. The MT influence network will be generated through a measure of indegree, which is the number of incoming ties. We are also interested in individuals (nodes) that have a high level of both types of relationships (RT/outdegree and MT/indegree), as this reflects the idea that they are highly influential in the network and are able to establish relationships with a large number of others in the network. One key aspect of this work therefore is to show how influence may be generated and captured in this social system through the identification of *superhubs* in big social systems as they engage around Twitter. These superhubs are individuals who are engaging in both high levels of RT and MT exchanges and represent "elite" influencers given their activity.

## Common Core State Standards (CCSS)

To present a new method that unites SMM, SNA, and Twitter it was necessary to choose a research focus that fulfilled two conditions: 1) to be the most significant as possible in its social repercussion without necessarily having a global scale, and 2) that the opinion climate was active through time. The CCSS provided the perfect vehicle to undertake such a research endeavor and is briefly described below.

## Social Context of CCSS

In the United States the existence of national norms is rare because traditionally each State establishes their own education standards. This has been changing because of the application of the CCSS. The CCSS in Math, Art and English were developed by the *National Governors Association* and the *Council of Chief State School Officers*, and have currently been adopted by 46 States. The CCSS have as the main objective to define the knowledge base required for students from Kindergarten to High School before getting into College. The CCSS require a re-orientation in each State and District, and in many cases, a profound revision of the curricula, teaching methods, and evaluations. As is, the CCSS imply a high-level change in education policy in terms of reach and depth as well as having the potential to impact policy, practice, and leadership.

As there is a dearth of studies around the CCSS, particularly from social systems perspective, we set out to undertake the following:

1. To represent the Complete Network and identify the percentage of it that represented the Giant Component (GC), meaning the overall systems of interaction.
2. To represent the structural communities existing within the GC.
3. To represent the network of the 1% of the Complete Network that reflected the superhubs (high influence) within these networks.
4. To represent the network of the 0. 25% of the Complete Network to identify those superhubs, who were the most active in these systems.
5. To represent the geolocalized social activity of the users, which indicates their physical location.

## Method

The collection and extraction of the data presented in this investigation was obtained directly from Twitter's API (*application programming interface*) during a six-month period, from September 1st 2013 to February 28th 2014. We used the *hashtags* "CommonCore" or "Common Core" to gather the 183,346 tweets during the time frame, from which 89,450 (48.7%) were identified as RTs and 129,339 as MTs (see Figure 9.1). We also generated the complete network (including all individuals) and giant component (network with isolates removed) from the data set as well. The goal is to represent this wider system and share a small part of the overall project.

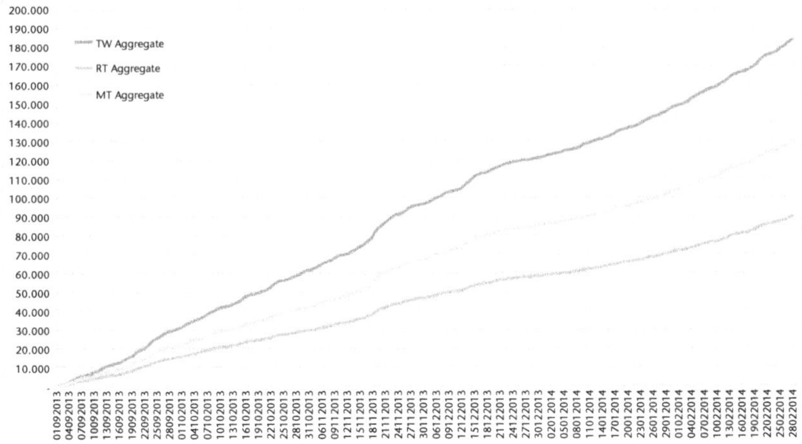


**FIGURE 9.1** Evolution of All Tweets, MTs and RTs over Six Months (September 2013-February 2014) from #commoncore

## Results

Using all of the data collected over the six-month period we were able to portray the social system around the CCSS. The total Complete Network (meaning everyone captured in the data) included 63,493 nodes (unique individuals) and 125,879 exchanges that result in the entire network (see Figure 9.2). The Giant Component (GC) (meaning the network with all isolates removed) resulted in 51,254 unique individuals (80.72% of the Complete Network total) and 123,035 relationships (97.74% of the Complete network total) (see Figure 9.3), which gives an idea of the activity of this social system over this six-month time period. For all of the graphs each node (dot) represents a unique individual that tweeted to the common core network and the lines between reflect exchanged tweets. Larger nodes reflect individuals that engaged in more activity (meaning sending/receiving more tweets).

Once we represented the structure of the complete network and the giant component we calculated several key metrics to gain a deeper understanding of the structure of the network. The key metrics we drew on were *modularity*, used to identify groups, *in-degree*, a measure of incoming ties/mentions (MT) per actor, and *out-degree*, a metric that measures the number of outgoing ties/re-tweets (RT) to other actors to arrive at the at the set of nodes (individuals) at the 1% and 0.25% level who reflected the most active individuals in this system who have disproportionate influence over the exchanges within the system.

We start with an analysis using the modularity metric on the GC to identify unique groups/sub-systems. These groups are defined by having more activity between group members than between groups based on their Twitter activity. This analysis resulted in three unique structural communities in which individuals

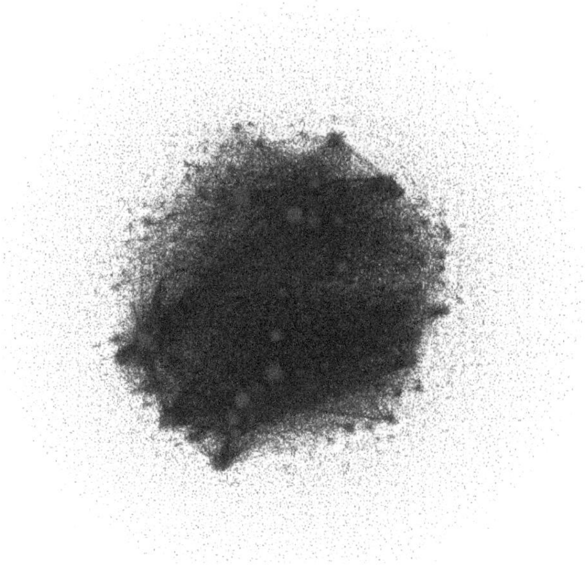

FIGURE 9.2 Complete Network of CCSS over Six Months.

FIGURE 9.3 Giant Component Network of CCSS over Six Months

engaged more with one another than those from another group (see Figure 9.4). This suggests that while there is a much larger interconnected system around the common core, there are also unique communities or sub-systems that are interacting within the larger system that have their own set of interactions.

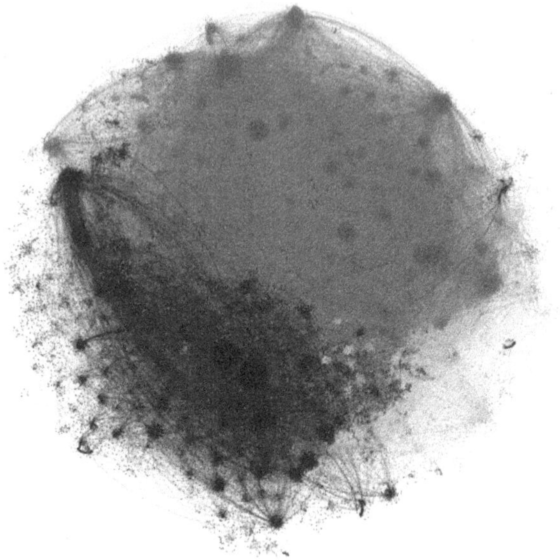

**FIGURE 9.4** Giant Component with Structural Communities.

Further analysis that is reported on the hashtagcommoncore.com website reveals that each of these communities reflected pro, con, and mixed views around the common core. It is important to note that we did not a priori determine these subsystems, rather they are based on "observing" the Twitter activity of these actors over the six-month period. We did not "impose" our structure upon the system, rather we just analyzed the actual behaviors from which a structural system arose.

Once we determined the overall structure of this large network of individuals and the subsystems we then wanted to better understand the system of interactions of those individuals who were most active in the system. In conducting this analysis, we focused at the 1% and .25% level to provide a drill down into the wider system. In order to establish these levels of analysis we filtered out all other actors to reach 1% of individuals within the greatest outdegree (RT) and indegree (MT) (as we define above) activity above the average actor in the network. We then filtered to the top .25% of the outdegree and indegree network to reveal the most highly active individuals who were over 2 standard deviations above the mean in their Twitter activity.

The 1% indegree (incoming ties/Mentions) network included 643 unique individuals (1.01% from total); 3,811 relationships (3.03% from total) (see Figure 9.5). Applying a *key player* procedure (Borgatti, 2006), which identifies highly central and influential actors in this network, we found 18 *superhub* nodes (2.79%) who could reach 94.5% of the network to reach *site percolation* (Newmann, 2010). *Site percolation* may have important effects in

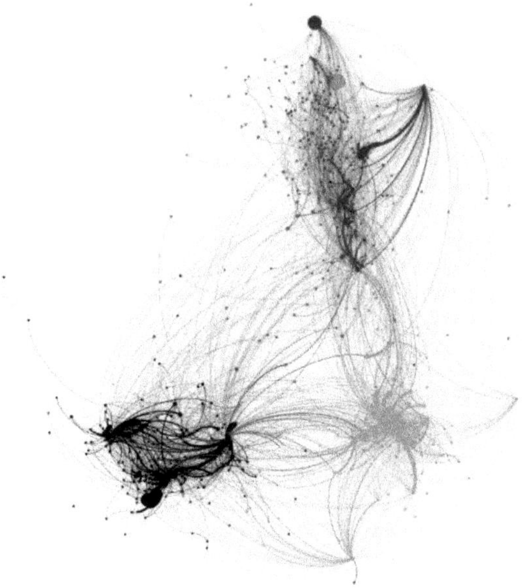

**FIGURE 9.5** 1% Common Core Network by Indegree/Mentions

the network as it is a way to stop or expand the movement of information, ideas and perceptions and may cause a chain effect within the network in terms of supporting or constraining the movement of resources (e.g. ideas, information). Thought of another way, only a few individuals had major influence on what was moving within the common core network and could both support or inhibit the movement of ideas/opinions – in essence reducing the "novelty" that was available to the system and having their opinions/ideas dominate the exchanges.

When we examined the 1% network by out-degree (meaning those actors that were sending disproportionate numbers of retweets into the system), we identified 641 nodes (1.0% from total) reflecting 3,140 relationships (2.49%) (see Figure 9.6). When we applied a *key player* metric to this network we found 19 *superhub* nodes (2.96%) who can access almost 95% of the network, which again is a very small fraction to achieve *site percolation*.

Although the 1% network was useful in understanding the overall structure of those individuals who were highly active in the system, we were also interested in the next level of influence, meaning those individuals who were in the top 0.25% of the overall network and reflecting 2 standard deviations above the mean of activity. The 0.25% network by in-degree/mentions included 163 nodes (0.25% from total) and 882 relationships (0.7% from total) (see Figure 9.7). We also applied the *key player procedure* to this network and found 9 *superhubs* that could reach 95.6%.

**FIGURE 9.6** 1% Network by Out-degree/Retweets

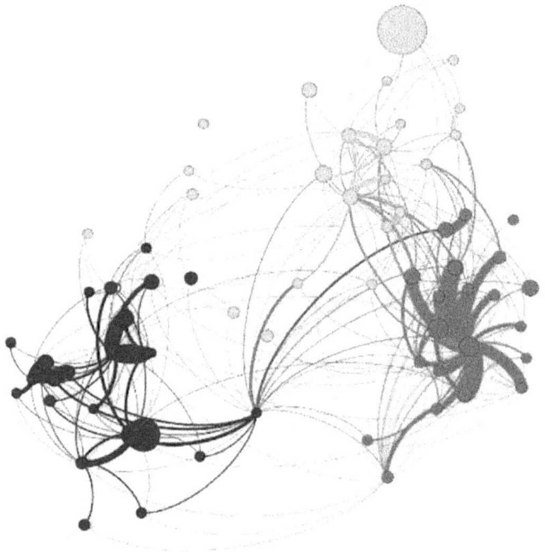

**FIGURE 9.7** 0.25% Network by Indegree/Mentions

The 0.25% network by out-degree/retweets included 158 nodes (0.25% from total) reflecting 705 relationships (0.56% from total), (see Figure 9.8). The *key player metric* resulted in the finding that 95.1% of the network could be reached with only 5 *superhub* nodes (3.1%), suggesting a very small elite group of actors that were dominating the conversation within the network.

**FIGURE 9.8** 0.25% Network by Out-degree/Retweets.

We were also interested in the geographic identification of the network as well. This algorithm restricts the analysis to those individuals who make their location public on Twitter and as such is limited by those that share location. In addition, we further restricted the results to the United States, which represented the vast majority of the locations for the common core network. Figure 9.9 portrays the indegree/mentions activity by location in the United States by the two major structure communities (red/pink and blue). As can be identified, the major concentration of activity was in the northeastern portion of the US, particularly around the New York area where the common core was hotly debated. This representation enables us to think about systems from the interaction between both social and physical space and as such opens up ideas about systems even wider.

We also analyzed Geolocalization by out-degree/retweet activity with the same restrictions as in the indegree network (see Figure 9.10). As in the indegree/mentions network the activity is highly concentrated on the east coast of the United States with a growing group of activity on the west and central part of the country.

## Discussion

Although it seems social media and activity on the net have been ever present, it is important to have some context. The last major federal education policy, No Child Left Behind (NCLB), was rolled out just three years after Google's founding. However, what is more extraordinary is that NCLB was enacted

**FIGURE 9.9** Complete Network. Geolocalization by In-degree/Mentions.

**FIGURE 9.10** Complete Network. Geolocalization by outdegree/retweets.

three years before Facebook was started; four years prior to YouTube, five years after Twitter. For the first time in education policy history social media is a player in a way that was unprecedented – and so while "media" has been around, it is the "social" modifier that captured our imaginations and caused us to consider systems in new ways in this chapter.

This research was undertaken to better understand the larger systems that surround policy makers, educators, and leaders. We also took a social perspective as our experience suggests that educational systems and actors within those systems are often treated as independent units rather than as a larger connected system. Typically, educational institutions and support agencies have not viewed themselves, either as organizations or individuals, as part of a larger interdependent and interconnected system or network. This failure to recognize and embrace the idea that decisions, actions, and inactions are mutually influential and consequential has perhaps inhibited our collective ability to both understand and address pressing issues that have for far too long plagued education.

The first finding from our work is that the common core network reflected a consistent social and active network of tweets and exchanges not only throughout the span of what we report in this chapter (six months), but well beyond. The overall high level of social activity around the CCSS was impressive, particularly given its consistency and longevity over the course of the six months of the project. Moreover, we were able to show the larger structure of this network in a way that has not been done previously.

We also found that within the common core network there were separate and overlapping structural communities. These sub-systems were based on the social activity of the actors and resulted in three distinct and overlapping groups. We did not "pre-assign" these factions a priori based on attributes of the individuals, rather we let their interaction activity on Twitter determine the group to which they belonged and then used that data as the starting point to identify actors within those communities. In this way we intersected the idea of activity and social space to better understand the interactions between and among those active in this system. Further analysis reported in the hashtagcommoncore.com website suggest that these factions shared similar belief patterns and at times were often echo-chambers. Another implication for this work is that even with larger systems there are subsystems which may seem like the entire "world" until one zooms out a bit further to show the larger system of interactions. A systems perspective from this stance requires both zooming in and out.

Evidence from this work suggests that the pattern of interactions captured in the activity of tweeters to #commoncore reflects a larger interdependent social network. Our findings indicate a high rate of interconnectivity and sharing of ideas, knowledge, and opinions from users on Twitter that created a massive social network. The total network included about 63,000 actors, 51,000 of whom interacted with multiple other actors over time, creating a dynamic set of activity exchanges across and between users. The network remained alive and grew even more dense over the course of the research, suggesting the robust nature of the network.

Our analysis identified influential actors or superhubs based on not only their Twitter activity, but interactions with others within the common core network creating a mix between activity and social space. This interaction created a system that was not based on self-report from individuals, but from a careful analysis of the interaction activity patterns within a specific social space. This analysis yielded a number of individuals who were highly active and influential in diffusing their beliefs and opinions. The network approach to the analysis enabled us to identify actors that both played an important role in this network and may be traditionally overlooked as they do not have a large number of followers, yet wield considerable influence. Our work illuminated subgroups and individuals that often are in the shadow, but who nonetheless played important roles in this new view of systems.

The sum of these four points reflects our thinking that "space" extends beyond physical geographies to encompass the idea of activity and social geographies to encompass a wider systems perspective. Further, portraying the sets of interactions surrounding a large-scale educational policy may enhance our understanding of such complex phenomena.

At its heart, the work presented in this chapter is based on the proposition that activity in social space produces complex systems that are situated in networks of interactions and interdependence. We live in a social world and as such are deeply affected by others, sometimes in ways of which we are unaware. Recent research suggests that our happiness, health, weight, and even wealth are influenced by the social network systems in which we reside (Christakis & Fowler, 2009). These connections are not just direct ones, but even those individuals who are two steps away from us have influence. However, despite this we still tend to draw on a variety of formal structures and processes to understand our worlds. However, while these more technical approaches are important and have been well documented, what appears to be generally missing is the power of the relational linkages between individuals that, while consequential, is often invisible. Therefore, examining both the quantity and quality of social ties between and among individuals in social media space is important and broadens our notions of system.

Our ability to actively interact with others, to network, and draw on systems perspective is of critical importance as we move deeper into a knowledge economy in which collaboration, social skills, and leveraging interdependent social networks are increasingly important, necessary, and hold potential economic/social/political/cultural value. Better understanding of newly evolving concepts and findings from network science and systems thinking are important for adding to our knowledge and building our own individual and collective ability to learn, lead, and understand processes of large scale policy implementation. This work suggests that the work of the twenty-first century is not only about facts, figures, and rote learning, it is about the generation of intellectual capital and the creation, development, and management of knowledge and opinion as it exists in multiple complex and dynamic systems. Approaching the work of education politics and policymaking as a system of relations recognizes that while the individual is important, it is the system of interactions amongst individuals that is equally informative. A social network perspective reveals the consequential interactions that are hidden in plain sight.

## References


Borgatti, S. P. (2006). Identifying sets of key players in a network. *Computational, Mathematical and Organizational Theory, 12*(1), 21–34.

Christakis, N. A., & Fowler, J. H. (2009). *Connected: The surprising power of our social networks and how they shape our lives*. New York: Little Brown.

Cross, R., & Parker, A. (2004). *The hidden power of social networks: Understanding what really gets done in organizations*. Cambridge, MA: Harvard Business School Press.

Daly, A. J. (2010). *Social network theory and educational change*. Cambridge, MA: Harvard Education Press.

del Fresno, M. (2014). Haciendo visible lo invisible: Visualización de la estructura de las relaciones en red en Twitter por medio del Análisis de Redes Sociales. *El profesional de la información, 23*(3), 246–252.

Dorsey, J. (2012). *Twitter* takes the pulse of the planet. It's the intersection of every media & medium. *Twitter*, November 15.

Guandong X., & Lin L. (ed.). (2013). *Social media mining and social network analysis: Emerging research*. Reference and Research Book News, 28(3). Portland: Ringgold Inc.

Granovetter, M. (1983). The strength of weak ties: A network theory revisited. *Sociological Theory, 1*(1), 201–233.

Granovetter, M. (1985). Economic action and social structure: The problem of embeddedness. *American Journal of Sociology, 91*, 481–510.

Jinhui, T., Meng, W., Xian-Sheng, H., & Tat-Seng, C. (2012). Social media mining and search. *Multimedia Tools and Applications, 56*, 1–7.

Moreno, J. L. (1934). *Who shall survive? Foundations of sociometry, group psychotherapy and sociodrama*. New York: Beacon House.

Newman, M. E. J. (2010). *Networks*. Oxford: Oxford University Press.

Scott, J. (2000). *Social network analysis*. London, UK: Sage Publications.

Supovitz, J., Daly, A., & del Fresno, M. (2015). *#commoncore Project*. Retrieved from http://www.hashtagcommoncore.com

Zafarani R., Abbasi, A., & Liu, H. (2014). *Social media mining*. Cambridge University Press.

# 10

# USING THE SYSTEMS THINKING APPROACH IN EDUCATIONAL POLICY SETTING

## A Choice Among Compromises

*Adam Nir*

Educational policy setting, traditionally considered a challenging and difficult task, has become nowadays complicated more than ever before. Much of the complexity follows the ongoing changes taking place in our environment and the increasing number of constituencies making competing claims regarding educational goals and processes based on evidence which best supports their interests (Lubienski, Scott, & DeBray, 2014).

While the articulation of policies in public education has some common denominators with policy setting processes conducted in other sectors, the features of public schooling present unique challenges and complexities for policy planners that are not easily managed. Not only are educational policy planners exposed to a variety of contradictory interests characterizing the turbulent environment in which public schools operate (Schon, 1983); they are also expected to define policy plans that offer solutions to ill-defined and unclear problems (Leithwood, 1994; Wildavsky, 1979) which may have multiple correct solutions and multiple methods of obtaining each (Wagner, 1994).

Educational policy planners perform their task in a social context overwhelmed by questionable counter-evidence promoted by various interest groups and stakeholders who attempt to challenge existing scientific truths (Oreskes & Conway, 2010; Specter, 2010). In addition, they need to cope with political ideologies and agendas creating a prominent source of stress (Blanco & Grier, 2009; Green, 1997; Lawton, 1992) which typically tends to increase before election time (Monchar, 1981; OECD, 2004; Paul, 1991; Popkewitz, 2000; Stevenson & Baker, 1991; World Bank, 2004). This last statement may be better understood in considering that educational policies are often expected to satisfy the contemporary political thought and governments' interests (Popkewitz, 2000) which tend to change over time (O'Toole, 1989).

As a result, educational policy planners are constantly exposed to a dilemma occurring between political interests and professional considerations. Consequently, the articulation of educational policy plans attempting to offer comprehensive solutions to educational needs and problems becomes a task exposing policy makers to conflicts among different truths and interests that may not be easily bridged or combined.

The point of departure for the current chapter is that policy setting in public educational systems is not merely a rational integration among a system's components qualities and goals. Rather, it is the act of choice among compromises, each representing a different combination of interests and values. In this sense, the systems thinking approach may be useful in considering its reputation in assisting management to confront situations characterized by dynamic change and complexity (Jackson, 2009; Jolly, 2015).

## The Systems Thinking Approach

The systems thinking approach advocates the need to see the world as a complex system where everything is connected to everything else (Sterman, 2000, p. 4). It emphasizes wholes rather than parts (Miller-Williams & Kritsonis, 2009) and the relationship among various components in a process, rather than the independent constituents of the process (Gregory & Miller, 2011). According to the systems thinking approach, the analysis and understanding of an organizational system should be based on a dual perspective that simultaneously looks at the whole system and at each of its components (Shaked & Schechter, 2013; 2014; 2016). Systems thinking is based on the theoretical premise stating that looking at the system as a whole is likely to produce better outcomes with fewer unintended consequences than analyzing and improving each element on its own. It enables better understanding of how systems work over time within the context of larger systems (Miller-Williams & Kritsonis, 2009). Employing the systems thinking approach is assumed to lead to greater understanding through integrating the act of analysis which answers the questions "what" and "how" and the act of synthesis which answers the "why" and "what for" questions (Laszlo, 2012).

The systems thinking approach has the potential to produce many benefits for both individuals and organizational systems. At the organizational level, it helps systems "to make the full patterns clearer" (Senge, 1990, p. 7), view events in a new light and respond to them in a genuine and innovative way. The systems thinking approach also benefits organizational change initiatives since it promotes comprehensive reforms rather than just piecemeal changes (King & Frick, 2000; Miller-Williams & Kritsonis, 2009).

At the individual level, systems thinking is considered to improve critical thinking capacities and decision making processes and skills (Chang, 2001; Lyneis & Fox-Melanson, 2002; Stave & Hopper, 2007). Systems thinking skills are viewed as significant in the process of problem solving as they allow individuals to change their own mental models, control their ways of thinking (Ben-Zvi Assaraf & Orion, 2005; Kim & Senge, 1994) and, therefore, produce better decisions.

Adopting a systems thinking approach seems to be of value in the context of schooling as well. Public educational systems expose school leaders to a realm characterized by massive complexities and ambiguity. Therefore, they often need to employ a different way of thinking about their organizations and search for new ways to handle the challenges they face (Miller-Williams & Kritsonis, 2009). The claim arguing that systems thinking may help to cope with these challenges is supported by empirical findings indicating that systems thinking helps educational leaders to establish the relations between schools and their communities for the benefit of their students (Chance, 2005), to develop schools as professional learning communities (Wells & Keane, 2008) and to reframe school–parents' relationships (Price-Mitchell, 2009). At the national level it is argued that systems thinking may help to re-conceptualize national policies as well as the various and interconnected consequences of their implementation (Gomez et al., 2015).

One of the key features characterizing organizational systems may be evident in the cause–effect relationships typically existing between parts of a system, as systems are composed of interconnected parts (Richmond, 1993; 1994; 1997). Therefore, understanding the way different components which make up a system are connected and the relationship between the whole system and each of its components appears to be crucial for the analysis and understanding of an organizational system as stated by the systems thinking approach.

While this premise seems to hold true for most organizational systems, the point of departure for the current chapter is that the systems thinking approach should be used with care when employed during policy setting processes performed in highly fragmented organizational systems, typically hosting contradictory interests, agendas and processes. In such organizational contexts, any attempt to produce comprehensive policy plans by looking at the sum-total of the system's various components may be misleading.

## The Unique Case of Loosely Coupled Systems

Schools and public educational systems are organizations characterized by loose coupling more than any other system (Fusarelli, 2002). This feature refers to the nature of reciprocal relations that exist among the components of a system (Weick, 1976). In loosely coupled systems, the various components maintain their autonomous and unique identity (Meyer & Rowan, 1977). Although they interact with each other, the relationships between them are characterized by low frequency and little dependency. These relations are multi-directional and may be formal or informal (Goldspink, 2007). Under loose coupling the degrees of freedom and autonomy characterizing each of the system's elements are high, limiting an attempt to describe and characterize a system using a linear rational and hierarchical perspective (Orton & Weick, 1990; Weick, 1976). Nevertheless, while processes of language and culture may be unique to each of the system's elements, they will not necessarily have a negative impact on the system's performance as a whole (Brusoni & Prencipe, 2001).

Loose coupling may be evident in public educational systems in the lack of dependency or binding relationships that exist between various processes taking place in different districts, schools and the national level. Loose coupling enables different stakeholders to offer different interpretations to the regulations and to the policies set by the state which may lead to inconsistencies. At the same time, however, loose coupling may increase schools' ability to cope with their typical daily challenges. These include interests characterized by contradictions, duplications, and gaps in coverage, pluralistic and sometimes mutually competing goals and a variety of elements that seldom mesh neatly with one another (Nir, 2000). In this sense, loose coupling may be viewed as a mechanism fostering schools' adaptability and relevancy to their local environment and their students' needs.

The loosely coupled nature of educational systems, the variety of interests involved and the lack of causal relations among the systems' components and processes create a unique challenge for system analysis and policy setting processes. In this respect, looking at educational systems as wholes as suggested by the systems thinking approach and analyzing the relationship between the whole system and each of its components while setting national policies may add to the inherent complexity typically characterizing these processes. In light of the fragmentation and internal variability characterizing educational systems, it may be understood why educational policy setting is seldom a rational integration among a system's qualities, goals, and professional considerations. Rather, this process is often an act of choice among compromises, each representing a different combination of qualities, aspects, and interests. Such a strategy enables people to better cope with the variety of interests involved and reduce stress and potential conflicts.

## Compromises in Educational Policy Setting: The Three Ps

Educational policy plans are usually the meeting point for a variety of expectations. Therefore, in the process of educational policy setting, compromises are inevitable in light of the contradictory interests that educational policies are expected to satisfy. This variety of inconsistent interests sets the grounds for a range of dilemmas for policy makers which vary in scope: some are broader and yet fundamental to the very nature of public education (Bradley & Taylor, 2002). Others are specific to particular areas within the educational realm (Shapiro & Stefkovich, 2000).

Compromises typically characterizing national educational policies may be categorized to three main types, each representing a different set of considerations and implications (Nir, 2016):

**Type A compromise – Progression**: A policy plan based on a Type A compromise will present a moderate discrepancy from some ideal goals or agendas. However, it will still encourage the introduction of significant changes and the advancement of professional goals which are considered substantial and beneficial by all stakeholders. A policy plan that follows this compromise will enable the advancement of individuals, schools and the educational system as a whole and has the potential to produce significant changes and worthwhile outcomes in the long run through the innovations and improvements it fosters.

**Type B compromise – Preservation**: A policy plan based on a Type B compromise is mainly oriented towards maintaining stability and avoiding potential conflicts. It typically promotes less controversial ideas which seem to offer some useful outcomes in light of the numerous contradictory interests and limitations involved. However, such compromises are somewhat misleading since a policy plan offering moderate changes and improvements of current patterns and processes is more likely to promote stagnation rather than the development of educational systems. Policy plans promoting a Type-B compromise may turn dysfunctional in the long run, especially if schools operate in turbulent circumstances characterized by massive and rapid changes.

**Type C compromise – Paradoxical**: Policy plans based on paradoxical compromises are typically articulated when circumstances are characterized by contradictory values and extreme expectations that do not have a point of convergence and, therefore, cannot be easily bridged. Paradoxical compromises may often follow the Golden Mean Fallacy asserting that extreme points of view must be wrong or less relevant and that a middle point between these extremes is always better and truer (Gula, 2002). Such compromises are self-contradicting and, therefore, plans that are based on paradoxical compromises tend to offer unproductive solutions likely to undermine the effectiveness and development of educational systems.

## The Israeli Educational Context: Some Basic Facts

The educational system has always been the primary vehicle for diminishing social gaps in the Israeli society and ensuring social, political, cultural, and economic mobility for all. The system is controlled by a strong central bureaucracy located in Jerusalem and is run by the Minister of Education and a Director General.

The Ministry of Education is judicially and politically responsible for the enactment of laws and the operation of the educational system. It sets national goals, tightly controls inputs and the allocation of budgets, monitors and controls student achievements through national performance evaluation tests, determines the national curriculum, and is responsible for employing teachers and the construction of new schools (Glasman, 1986). The educational system is divided into seven districts which supervise and monitor the educational processes conducted by schools to ensure the compatibility of these processes with central policies.

In the school-year 2015, the Israeli educational system included 4,972 schools (The Knesset Report, 2015) and 2,162,628 pupils (Central Bureau of Statistics, 2016). The official state school system consists of two trends in accordance with the "Law of State Education" enacted in 1953 recognizing each religious sector's right to autonomy regarding separate institutions, teachers, inspectors, and curricula (Gaziel, 1996): "State Schools" serving the non-religious sector and "State Religious Schools" serving those seeking greater emphasis on religious Judaism in the curriculum. The ratio of enrollment in the two trends is about 70% in the state system and 30% in the state-religious system. Two additional sectors whose schools belong to the official school system are Israeli-Arabs and Druze where teaching is almost exclusively conducted in Arabic. Christian-Arabs and Ultra-Orthodox Jews are two additional sectors receiving educational services in exempted schools outside the state school system. This group of schools consists of Arab Mission schools supervised and controlled by the Catholic Church, and Jewish traditional Talmud-Torahs supervised by Ultra-Orthodox political parties and the rabbis of the Ultra-Orthodox community.

The public school structure is divided into four main stages: kindergarten (from age 3 to 6 years), elementary school (grades 1 through 6), middle school (grades 7 through 9), and high school (grades 10 through 12). The "Compulsory Education Law" also allows parents to have their children educated according to their religious views (Bentwich, 1965). Therefore, children may enroll in one of three groups of schools: (a) "official" (i.e., maintained by the government and the Local Education Authorities, or LEA), (b) "recognized" (i.e., supervised and supported by the government), or (c) "exempted" (not seeking government supervision or support and in practice subjected to no government control). High schools are run either by LEAs or by semi-public/private non-profit organizations.

Although great diversity exists among the different LEAs as a result of economic and political constraints and geographical location, academic standards are controlled through national matriculation examinations conducted by the Ministry of Education. In elementary schools, in addition to school-level efforts to track student achievement, control is obtained through national evaluation tests.

In the last few decades, however, the Ministry of Education initiated various polices intended to decentralize rather than centralize its control patterns. These initiatives followed pressures to increase schools' autonomy based on the recognition that strong centralization and curriculum uniformity have a negative pedagogical effect on schools (Vollansky & Bar-Elli, 1995) and may undermine the relevancy of educational processes (Nir & Eyal, 2008).

## The Israeli National Task Force for the Advancement of Education and School Autonomy: A Case Analysis

The usefulness of compromise analysis in relation to educational policy setting may be demonstrated by looking at the policy plan articulated by the Israeli National Task Force for the Advancement of Education. This Task Force was appointed in 2003 by Limor Livnat, Minister of Education, Culture and Sports, with the support of Prime Minister Ariel Sharon and Finance Minister Binyamin Netanyahu. The Task Force was charged with conducting a comprehensive examination of the Israeli educational system and recommending an inclusive plan for change – pedagogical, structural and organizational – as well as outlining the means for implementing it. The Task Force formed 12 professional committees made up of more than 100 professionals, each addressing a different major topic. After 15 months of work, the committees submitted their recommendations to the Task Force which compiled an intermediate report (The State of Israel, 2004). A final report was published a few months later. The final report presents an analysis of the Israeli educational system, a proposed vision, and a detailed description of the national policy plan with an outline for its implementation (The State of Israel, 2005).

The appointment of the National Task Force took place approximately six years after the Ministry of Education articulated and began the implementation of a School-Based Management (SBM) policy. This initiative followed past unsuccessful efforts to decentralize the Israeli educational system and increase schools' autonomy. It is worth mentioning that previous initiatives had little effect on the centralized nature of the Israeli educational system, mainly because of the "centralization trap" (Nir, 2006) evident in the contradictory tendencies to delegate authority to schools and, at the same time, to maintain substantial central control in schools (Nir, 2003a; 2003b). The Task Force, even so, decided to embrace the SBM policy along with its guiding assumptions in recognizing the benefits embedded in school empowerment for students, local communities and the educational system as a whole. Nevertheless, an analysis of the Task Force recommendations concerning school autonomy reveals a variety of compromises that follow an inclination to reconcile between these contradictory tendencies.

## First Example: The Glass Ceiling for School Empowerment

A basic assumption in the proposed reform is that schools' autonomy must be embraced and strengthened:

> Strengthening the status and autonomy of school – **Maximum** pedagogical, administrative and budgetary autonomy **for the schools**, led by principals as the leaders of the educational activity, will enable the schools to realize their commitment to high-quality educational activity.
>
> *(The State of Israel, 2004,* Intermediate Report, *p. 9)*

> We highly value significant improvements in the nature and performance of schools and kindergartens and the **significant empowerment** of those who are at the center of the educational activity.
>
> *(The State of Israel, 2005,* Final Report, *p. 14)*

> **Maximum** pedagogical, administrative and budgetary **autonomy** for the schools, led by the principals as the leaders of the educational activity, will enable the schools to realize their commitment to high-quality educational activity. . . the school shall select its employees.
>
> *(*Final Report, *p. 15)*

These statements offering to increase schools' autonomy to the maximum seem to be in line with the assumptions characterizing a Type-A compromise. They indicate that policy makers consider school autonomy an ideal which has to be adopted in light of its potential to advance the educational system and, therefore, to strive to promote the autonomy of schools. However, can we conclude based on these statements that policy makers actually intended to empower schools by handing school-level educators maximum authority for the various processes conducted in their schools? Other policy messages suggest differently. One expression to the limited degrees of freedom teachers and school principals are likely to obtain may be found in statements referring to their ability to shape their school goals according to their educational agenda and the needs of the local community they serve:

> The Ministry of Education will set measurable indicators for the entire system and will update them every three years.
>
> *(*Intermediate Report, *p. 12)*

> The Ministry of Education will determine, with the approval of the National Council for Education, measurable objectives for the entire educational system. These objectives will reflect the priorities and policy

of the Ministry of Education. These objectives will be set for a period of five years and will be updated every year. They will serve as the annual objectives for the educational system.

(Final Report, *p. 165)*

. . . under the new organization, the Ministry of Education will become a body which sets policies, finances, sets standards and ensures that they are carried out.

(Final Report, *p. 28)*

Hence, along with the messages emphasizing the significance of school autonomy, other messages emphasize the central role of the Ministry of Education as the body which determines goals, priorities and performance standards. Other messages further reflect the restrictions on school principals' authority:

. . . teachers' work conditions and salaries will be set according to **collective agreements**. Setting special contracts will be allowed only in special cases.

*(Intermediate Report, p. 11; Final Report, p. 23)*

School will be granted full flexibility in running its budget **provided its conduct meets the obligatory policy** and assignments set to the school

(Intermediate Report, *p. 55)*

School will have autonomy in using its budget **in the frame of the obligatory policy** set by the Ministry of Education and the regional educational administration

(Final Report, *p. 78)*

. . . **most** of the authority in the areas of pedagogy, budgets and personnel will be transferred to schools.

(Intermediate Report, *p. 10)*

Schools and kindergartens are at the center of the educational activity and they are the ones who are responsible for the education in the state. As an expression of trust in their ability and as means for their improvement, **most** of the authority and resources should be transferred to school principals and to school and kindergarten teachers.

(Final Report, *pp. 23; 78)*

It is evident that the commitment to transfer **full** authority to the school level as stated in the committee's Intermediate Report is moderated. The final report states that **most** authorities will be transferred to schools which will be subjected to various requirements set by the Ministry of Education and the regional educational administration. Hence, such statements seem to be in line with a Type-B compromise indicating the tendency to avoid potential conflicts and major changes by introducing limited improvements in current patterns of conduct.

### *Second Example: Boundaries of Flexibility*

Another principle emphasized in the report which is central to school empowerment involves the proximity between decision makers and local circumstances as means to increase the relevancy of decisions to local expectations and needs. This principle is clearly stated in both the intermediate and the final report:

> . . . schools need to be empowered . . . based on evidence coming from many studies showing that granting schools wide authorities allow decision making processes to take into account students' particular needs and promote educational achievements.
>
> (Intermediate Report, *p. 32;* Final Report, *p. 53)*

Did policy makers involved in the articulation of the recommendations ensure that school principals are left with sufficient authority, allowing them to design their school's pedagogy based on their professional judgment and local needs? A deeper look at some of the messages that appear in the committee's intermediate and the final reports suggests that this principle is only partially maintained:

> . . . the school administration has the autonomy to determine the organization of studies and to set the educational and pedagogical modes of operation in accordance with students' needs **as long as the obligatory policy is kept**, both the national and the local one.
>
> (Intermediate Report, *p. 53)*

> . . . the educational institution is responsible **to meet the objectives set for it** and for the achievements of its students.
>
> (Final Report, *pp. 40, 70)*

It appears that schools' ability to make decisions that are mostly influenced by their students' needs is rather limited since schools are obliged to follow a core national curriculum as means for obtaining their budgets:

> . . . the school curriculum will be based on the comprehensive **national** curriculum which includes a **core** curriculum and other programs allowing deepening and expansion in particular areas that may be selected by schools.
>
> (Final Report, *p. 83)*

Moreover, schools may encounter difficulties in conducting differential programs inspired by local needs since they are forced to conduct pedagogical and didactical activities that are based on obligatory national standards:

> . . . **one obligatory standard for all** derived from the core curriculum: what every student should know, understand and be able to perform.
>
> (Intermediate Report, *p. 48)*

> . . . **standards for achievements and expected students' skills will be defined** according to the core curriculum and according to the entire curriculum for all educational stages.
>
> (Final Report, *p. 22)*

> . . . we are not hesitant **to set clear objectives for achievements and accomplishments** of the educational system that will be anchored by public standards and used to test the entire educational system and each of its components.
>
> (Final Report, *p. 58)*

In fact, the policy plan defined by the Task Force grants school principals the freedom to conduct unique educational processes only after they meet the demands set by the obligatory national curriculum:

> . . . the school principal as the pedagogical leader of his school is responsible together with the entire administrative staff to decide, in collaboration with the school management and pedagogical council, what complementary educational programs will be taught in school in addition to the core curriculum.
>
> (Final Report, *p. 116)*

It appears that the obligatory curriculum and standards leave school principals the freedom to choose the teaching methods to be employed in their school provided that students' achievements are in line with the standards set and determined by the centralized Ministry of Education:

> . . . it is possible to force the school to follow the educational policy and meet the obligatory standards but it is important to limit intervention as much as possible and allow schools to decide how these objectives will be met.
>
> (Final Report, *p. 111)*

Although the report stresses the significance of flexibility and schools' ability to make decisions based on local needs, schools are required to act in accordance with an obligatory core curriculum and meet central standards which significantly narrow their degrees of freedom. Schools are left with rather limited authority to determine how processes intended to meet these centrally determined objectives will be designed and conducted. This combination of contradictory regulations may testify again for a Type-B compromise evident in policy makers' tendency to initiate a new policy and maintain current modes of operation at the same time.

### *Third Example: Unifying the Different*

One of the main claims in favor of the tendency to decentralize educational systems argues that empowered schools are better able to address variance and meet the different needs of students and the various expectations coming from members of their community. This quality was not overlooked by members of the National Task Force who highlighted the significance of diverse educational processes in a democratic society such as the Israeli one, which continuously absorbs immigrants and is comprised of many social groups characterized by different cultural backgrounds:

> Nationalism and pluralism: being a democratic country, the Israeli educational system should give expression to the variety of cultures, ethnic groups and streams comprising the Israeli society.
>
> (Intermediate Report, *p. 34;* Final Report, *p. 56)*

The reports further stress that the educational system is expected to assist in

> . . . closing the gaps among different parts of the society, lowering the walls between different social groups and promoting solidarity and social cohesiveness among all Israeli citizens . . . all this should be accomplished without discrimination or obliteration of unique views and perspectives.
>
> (Intermediate Report, *p. 35;* Final Report, *p. 56)*

Yet, along with the messages emphasizing the importance to maintain and preserve the unique features of the different social groups which comprise the Israeli society, the report also brings quite different messages, creating doubt regarding the potential of the schooling system to give voice to the unique and the different:

> . . . in the entire world, the blessing embedded in liberalism and in educational approaches that are less strict **has its price which is not a simple one**
>
> (Intermediate Report, *p. 26;* Final Report, *p. 47)*

> Israel is a heterogenic society . . . we estimate that the large number of sectors and streams in education **is a central problem** for the educational system, being a divisive factor in the social-cultural tissue; it causes the loss of solidarity and partnership in the Israeli society and is a source for a large waste of resources.
>
> (Intermediate Report, *p. 28;* Final Report, *p. 49)*

To avoid paying this "not simple" price and the "large waste of resources," the reports offer to create a core curriculum:

> . . . we recommend expanding the core curriculum by adding clear definitions of contents based on educational objectives.
>
> (Intermediate Report, *p. 47)*

> . . . the core curriculum is a consolidating basis for all the different streams in the country and it creates a common denominator for all students both in the conceptions, contents and values and in their thinking and learning skills.
>
> (Final Report, *p. 86)*

> . . . clearer definitions of contents that are based on the educational objectives which determine what each graduate of the Israeli educational system must know and understand will be added [to the core curriculum].
>
> (Final Report, *p. 87)*

These statements indicate a paradoxical type of compromise. They show that in spite of the declarations emphasizing the need to serve students coming from different social and cultural backgrounds, the policy plan fosters a core curriculum and unified standards, which pay little attention to the unique and the different.

## Concluding Remarks

All three examples give evidence to the inherent complexity typically characterizing educational policy setting and the introduction of school autonomy policies in centralized structures. Specifically, policy planners need to cope with two contradictory agendas – promoting the autonomy of schools and maintaining strong centralized control over schools. Such circumstances promote policy makers' inclination to produce compromises and articulate policy plans that average these extreme perspectives to various degrees. However, paradoxical compromises produce paradoxical patterns of autonomy evident in the decelerations supporting school autonomy along with the articulation of measures intended to maintain central control (Nir, 2009). Such policies are likely to have limited positive effects and are less likely to promote the quality of educational processes performed in schools. They may also undermine school-level educators' trust in central initiatives and in policy makers holding positions at the Ministry of Education (Nir, 2000; Nir & Eyal, 2003). In summary, the use of compromises in the process of educational policy setting appears inevitable in light of the variety of contradictory interests involved. This mode of operation seems to meet the fragmented and loosely coupled nature of public educational systems fostering a variety of interpretations to any articulated policy.

When considering the fragmented nature of public educational systems and the variability of interests involved, it may be understood why educational policy-setting processes are so complicated and why the use of compromises is essential. Compromises moderate potential conflicts and decrease the stress which follows a clash between contradictory values. Articulating a policy plan which allows fulfilling all clashing interests seems almost impossible. In this sense, the systems thinking approach offers an important tool enabling to better analyze and understand the complexities involved (Jackson, 2009; Jolly, 2015). It may assist policy makers to re-conceptualize national policies as well as the various consequences of their implementation (Gomez et al., 2015) and to define more clearly the various possible compromises that may be employed for the benefit of schools and the entire educational system.

Taking into account the consideration to introduce school autonomy policies in centralized structures, a systems thinking approach may assist in detection of the main contradictory values and interests involved. This may help policy makers to articulate the best possible compromises likely to allow schools to take advantage of their increased autonomy and develop a genuine agenda while enabling central officials to sustain some control over schools.

Redefining school superintendents' formal role duties may illustrate a compromise along these lines. In considering that "even when increasing the independence of educational institutions, there is a need for the supervision of the superintendent" (Nir & Eyal, 2003, p. 551), promoting superintendents' involvement in capacity building activities and professional development processes provided for school level educators and decreasing their regulation and monitoring responsibilities, may create a realistic feasible and productive compromise. Changing the proportion between professional and bureaucratic role duties will transform superintendents into pedagogical leaders who are mostly involved in encouraging and assisting schools to promote new initiatives while ensuring that school processes are in line with central regulations. Such a compromise will enable reconciliation between the clashing interests evident in the tendency to simultaneously release and maintain central control over schools and still may produce worthwhile outcomes for schools and the educational system.

However, due to the complexity and fragmentation of public educational systems, a systems thinking approach should be employed within reasonable boundaries that limit the scope of the analysis. Such boundaries are needed to enable policy makers to see the big picture, and yet not fall into an endless detail-analysis and lack of congruence trap following the fragmented nature of public educational systems.

In spite of the power embedded in the systems thinking approach, it is important to acknowledge that any attempt to perform a comprehensive analysis of a loosely coupled system will only be able to produce a vague proxy of the countless interests and processes taking place in the system. Therefore, setting reasonable boundaries for system analysis processes is essential if policy makers are expected to choose feasible and relevant compromises and articulate policies likely to benefit the educational system.

The premise, therefore, stating that systems should be considered as wholes rather than parts (Miller-Williams & Kritsonis, 2009) as suggested by the systems thinking approach should be embraced with care in the process educational policy setting. Expectations regarding its potential to promote the quality and effectiveness of educational policies and increase the systems' internal congruence need to be adjusted accordingly.

## References

Bentwich, J. S. (1965). *Education in Israel*. London, UK: Routledge & Kegan Paul.

Ben-Zvi Assaraf, O., & Orion, N. (2005). Development of systems thinking skills in the context of earth system education. *Journal of Research in Science Teaching*, *42*(5), 518–560.

Blanco, L., & Grier, R. (2009). Long life democracy: The determinants of political instability in Latin America. *Journal of Development Studies*, *45*(1), 76–95.

Bradley, S., & Taylor, J. (2002). The effect of the quasi-market on the efficiency-equity trade-off in the secondary school sector. *Bulletin of Economic Research*, *54*(3), 295–314.

Brusoni, S., & Prencipe, A. (2001). Managing knowledge in loosely coupled networks: Exploring the links between product and knowledge. *Journal of Management Studies*, *38*(7), 1019–1035.

Central Bureau of Statistics (2016). *A forecast of students' numbers in the educational system for the years 2017–2021: Report, methodology and findings*. Jerusalem, Israel: Central Bureau of Statistics.

Chance, P. L. (2005). Engaging communities through vision development: A systems approach to public relations. *Journal of School Public Relations*, *26*(2), 139–155.

Chang, C. Y. (2001). Comparing the impacts of a problem-based computer-assisted instruction and the direct-interactive teaching method on student science achievement. *Journal of Science Education and Technology*, *10*(2), 147–153.

Fusarelli, L. D. (2002). Tightly coupled policy in loosely coupled systems: Institutional capacity and organizational change. *Journal of Educational Administration*, *40*(6), 561–575.

Gaziel, H. (1996). *Politics and policy making in Israel's education system*. Brighton, UK: Sussex Academic Press.

Glasman, N. S. (1986). Funding and control linkages in education. In E. Ben-Baruch & Y. Newman (Eds.), *Educational administration and policy making: The case of Israel*. Tel Aviv: Unipress (Hebrew).

Goldspink, C. (2007). Rethinking educational reform: A loosely coupled and complex systems perspective. *Educational Management Administration and Leadership, 35*(1), 27–50.

Gomez, R. E., Kagan, S. L., & Fox, E. A. (2015). Professional development of the early childhood educational teaching workforce in the United States: An overview. *Professional Development in Education, 41*(2), 169–186.

Green, A. (1997). *Education, globalization and nation state*. Hampshire, UK: Macmillan Press.

Gregory, A., & Miller, S. (2011, April). *Waving or drowning? Re-evaluating the place of systems thinking in business and management learning and curricula*. Presented at the Organizational Learning, Knowledge and Capabilities Conference, Hull, UK.

Gula, R. J. (2002). *Nonsense: A handbook of logical fallacies*. Mount Jackson, VA: Axios Press.

Jackson, M. C. (2009). Fifty years of systems thinking for management. *Journal of the Operational Research Society, 60*, S24–S32.

Jolly, R. (2015). *Systems thinking for business: Capitalize on structures hidden in plain sight*. Portland, OR: Systems Solutions.

Kim, D. H., & Senge, P. M. (1994). Putting systems thinking into practice. *Systems Dynamics Review, 10*(2/3), 277–290.

King, K. S., & Frick, T. (2000). *Transforming education: Case studies in systems thinking*. Indiana University website. Retrieved from https://www.indiana.edu/~tedfrick/aera99/transform.html

Laszlo, K. C. (2012). From systems thinking to systems being: The embodiment of evolutionary leadership. *Journal of Organizational Transformation & Social change, 9*(2), 95–108.

Lawton, D. (1992). *Education and politics in the 1990s: Conflict or consensus?* London, UK: The Falmer Press.

Leithwood, K. (1994). Leadership for school restructuring. *Educational Administration Quarterly, 30*(4), 498–518.

Lubienski, C., Scott, J., & DeBray, E. (2014). The politics of research production, promotion, and utilization in educational policy. *Educational Policy, 28*(2), 131–144.

Lyneis, D. A., & Fox-Melanson, D. (2002). The challenge of infusing system dynamics into a K-8 curriculum. *The Creative Learning Exchange, 11*(2), 1–9.

Meyer, J., & Rowan, B. (1977). Institutionalized organizations: Formal structure as myth and ceremony. *American Journal of Sociology, 83*(2), 340–363.

Miller-Williams, S. L., & Kritsonis, W. (2009). *A systems approach to comprehensive school reform: Using the realms of meaning and the Baldridge Model as a systems framework*. National Forum of Applied Educational Research Journal, *23*(1–2). Retrieved from http://files.eric.ed.gov/fulltext/ED506957.pdf

Monchar, P. H. (1981). Regional educational inequality and political instability. *Comparative Education Review, 25*(1), 1–12.

Nir, A. E. (2000). The annual plans of school-based management schools operating in a centralized educational system: Planning for ambiguity. *Educational Planning, 12*(4), 19–38.

Nir, A. E. (2003a). The impact of school-based management on supervision instructors' professional considerations. *International Journal of Educational Management, 17*(2), 49–58.

Nir, A. E. (2003b). The impact of school-based management on public schools in Israel. *Curriculum and Teaching, 18*(1), 65–80.

Nir, A. E. (2006). School empowerment and the centralization trap. In D. Inbar (Ed.), *Towards educational revolution?* (pp. 296–309). Tel Aviv: The Van Leer Jerusalem Institute and Hakibbutz Hameuchad Publishing House (Hebrew).

Nir, A. E. (Ed.) (2009). *Centralization and school empowerment: From rhetoric to practice*. New York, NY: Nova Science Publishers.

Nir, A. E. (2016). Educational planning: The ethics of compromise. *Educational Planning, 23*(1), 5–17.

Nir, A. E., & Eyal, O. (2003). School-based management and the role conflict of the school superintendent. *Journal of Educational Administration, 41*(5), 547–564.

Nir, A. E., & Eyal, O. (2008). Setting a national curriculum: Bridging the gap between professionalism and politics. In J. N, Casey & R. E. Upton, (Eds.), *Educational curricula: Development and evaluation*. (pp. 235–251). Hauppauge, NY: Nova Science Publishers.

OECD (2004). *Education policy analysis, 2004*. Paris, France: OECD.

Oreskes, N., & Conway, E. M. (2010). *Merchants of doubt: How a handful of scientists obscured the truth on issues from tobacco smoke to global warming*. New York, NY: Bloomsbury Press.

Orton, D., & K. Weick, (1990). Loosely coupled systems: A Reconceptualization. *Academy of Management Review, 15*(2), 203–223.

O'Toole, Jr. (1989). The public administrator's role in setting the policy agenda. In J. L. Perry (Ed.), *Handbook of public administration* (pp. 225–236). San Francisco, CA: Jossey-Bass.

Paul, G. (1991). *Schooling, skill, and the return of government investment in education: An exploration using data from Ghana*. New York: World Bank.

Popkewitz, T. S. (2000). *Educational knowledge: Changing relationship between the state, civil society, and the educational community*. New York: State University Press.

Price-Mitchell, M. (2009). Boundary dynamics: Implications for building parent-school partnerships. *School Community Journal, 19*(2), 9–26.

Richmond, B. (1993). Systems thinking: Critical thinking skills for the 1990s and beyond. *System Dynamics Review, 9*(2), 113–133.

Richmond, B. (1994). Systems thinking/system dynamics: Let's just get on with it. *System Dynamics Review, 10*(2/3), 135–157.

Richmond, B. (1997). The thinking in systems thinking: How can we make it easier to master? *The Systems Thinker, 8*(2), 1–5.

Schon, D. A. (1983). *The reflective practitioner*. New York: Basic Books.

Senge, P. M. (1990). *The fifth discipline: The art and practice of the learning organization*. New York: Currency Doubleday.

Shaked, H., & Schechter, C. (2013). Seeing wholes: The concept of systems thinking and its implementation in school leadership. *International Review of Education, 59*(6), 771–791.

Shaked, H., & Schechter, C. (2014). Systems school leadership: Exploring an emerging construct. *Journal of Educational Administration, 52*(6), 792–811.

Shaked, H., & Schechter, C. (2016). Sources of systems thinking in school leadership. *Journal of School Leadership, 26*(3), 468–494.

Shapiro, J., & Stefkovich, J. (2000). *Ethical leadership and decision making in education*. Mahwah, NJ: Lawrence Erlbaum Associates.

Specter, M. (2010). *Denialism: How irrational thinking hinders scientific progress, harms the planet, and threatens our lives*. New York: Penguin Press.

Stave, K., & Hopper, M. (2007, August). *What constitutes systems thinking?* A proposed taxonomy. Proceedings of the 25th International Conference of the System Dynamics Society. Boston, MA. Retrieved from http://www.systemdynamics.org/conferences/2007/proceed/index.htm

Sterman, J. D. (2000). *Business dynamics: Systems thinking and modeling for a complex world.* Boston, MA: McGraw-Hill.

Stevenson, D. L., & Baker, D. P. (1991). State control of the curriculum and classroom instruction. *Sociology of Education, 64*(1), 1–10.

The Knesset Report (2015). *The Israeli educational system: Some main issues under the jurisdiction of the Education Culture and Sports Committee.* Jerusalem, Israel: The Knesset Research and Information Center.

The State of Israel (2004 May). *The Israeli Task Force: Intermediate report.* Jerusalem, Israel: The Ministry of Education.

The State of Israel (2005 January). *The Israeli Task Force: Final report.* Retrieved from https://www.knesset.gov.il/docs/eng/dovrat_report_eng.doc

Vollansky, A., & Bar-Elli, D. (1995). Moving toward equitable school-based management. *Educational Leadership, 53*(4), 60–62.

Wagner, R. K. (1994). Practical problem-solving. In P. Hallinger, K. Leithwood, & J. Murphy (Eds.): *Cognitive perspectives on educational leadership* (pp. 88–102). New York: Teachers College Press.

Weick, K. E. (1976). Educational organizations as loosely coupled systems. *Administrative Science Quarterly, 21*(1), 1–19.

Wells, C., & Keane, W. G. (2008). Building capacity for professional learning communities through a systems approach: A toolbox for superintendents. *AASA Journal of Scholarship & Practice, 4*(4), 24–32.

Wildavsky, A. (1979). *Speaking truth to power: The art and craft of policy analysis.* Boston, MA: Little, Brown & Co.

World Bank (2004). *World development report, 2004: Making services work for poor people.* Washington, DC: World Bank and Oxford University Press.

# 11

# SCHOOL TURNAROUND REFORM

## Optimizing Confluence of Influence and Dynamic Disequilibrium

*Andrea Rorrer, Vicki Park, Cori Groth, and Janice Bradley*

The United States has had a long, and at times problematic, history of reform attempts aimed at improving low-performing schools and mitigating unequal access and outcomes for diverse student populations. School turnaround is one of those reform efforts. Having gained prominence since 2001 in the educational nomenclature and policy, the concept of "turnaround" has reflected a technical-structural approach to educational improvement. Of course, the resurgence of interest in turnaround was aided by rising demands for accountability and increased student performance. For example, turnaround appeared prominently in the two most recent iterations of the Elementary and Secondary Education Act (1965): No Child Left Behind (2002) and now Every Student Succeeds Act (2015).

In the midst of the federal and state policy changes, there have been multiple springboards for turnaround to further its impact, including the initiation of the American Reinvestment and Recovery Act (ARRA) (2009) and the Race to the Top (RTTT) (2009) grants and additional funding invested in the School Improvement Grants (SIGs). In the SIG Program, the turnaround model was one of four interventions for whole school improvement aimed at "persistently lowest-achieving schools" (U.S. Department of Education, 2009, p. 10). The SIG program promoted specific practices around four key areas: (1) adopting comprehensive instructional reform strategies; (2) developing and increasing teacher and principal effectiveness; (3) increasing learning time and creating community-oriented schools; and (4) allowing operational flexibility and support (Dragoset et al., 2017). During the past two decades, the connotations of turnaround have evolved. Ultimately, the federal turnaround model required replacing the principal and no less than 50 percent of the staff, in addition to implementing instructional reforms and increasing professional development and technical support (Dragoset et al., 2017). Now, we witness that turnaround has moved to being a more generic term encompassing, or synonymous with, school improvement and changes in outcomes.

For context, this study emerged from the work of the Utah Education Policy Center (UEPC) Bridgeworks team. The Bridgeworks team is a unit within the UEPC that works directly with schools, districts, and the state office of education specifically around building leadership for turnaround and school improvement. The UEPC approach to working with the educational system is founded upon a complex adaptive systems approach. This approach, which is grounded in the work of Plsek and Greenhalgh (2001), assumes the system is "a collection of individual agents with freedom to act in ways that are not always totally predictable, and whose actions are interconnected so that one agent's actions changes the context for other agents" (p. 625). Our research on school turnaround has previously addressed the process of reclaiming turnaround (Park, Groth, Bradley, & Rorrer, forthcoming). In this research, we highlight that:

> Reclaiming turnaround requires an active choice to move forward and away from a market-based, technical-oriented, and deficit approach to improving schools to one that places teaching, learning, and equity at the center. Building capacity to engage in school improvement for turnaround requires leaders, researchers, and partners alike to be willing to develop and learn moves that empower through co-construction, create shared meaning and visioning, cultivate continuous improvement and reflection, and facilitate assets-based learning approaches.

This chapter extends our scholarship on turnaround into how we begin to build longevity and impact of turnaround efforts. To this end, we reference turnaround as school improvement and school reform that carries the expectations for dramatic shifts in student achievement, including, but not limited to, achievement that is at or above state averages for all student groups. Specifically, we explore systems theory as both an analytical tool and a strategy for school turnaround reform. Thus, instead of foregrounding subsystem levels as the starting point of analysis or transformation, we center school turnaround as the unit of analysis. In doing so, we are able to consider how multiple subsystems work together in turnaround to achieve the reform intent. This enables us to focus on the processes and interactional relationships for influence as well as the leverage sources for change in turnaround reform. First, we describe the conditions that are precursors to taking a systems approach to examining turnaround as school reform. Next, we explain the importance of considering a dynamical systems approach with expectations for nonlinearity and variable coupling. Then, we present a typology of the interactional relationships for influence within the system based on three key dimensions: directionality of influence and change; degree of reciprocity; and primary interactional style.

Finally, we apply systems theory to the case of school turnaround and illustrate how a multi-dimensional sphere of influence can support a deeper understanding of implementing sustainable and scalable turnaround reform.

## Possibilities for Turnaround through Dynamical Systems Theory

Systems theory is an important perspective that emphasizes interrelated aspects, interdependencies between components, and patterns of change (see Shaked & Schechter, 2013). Rooted in the philosophical tenets outlined by Bertalanffy (1972), the core of systems theory is studying and understanding a system as a whole. Importantly, as Bertalanffy (1972) previously explained, systems thinking remains a rebuttal to existing traditional scientific approaches which emphasize "analytical, mechanistic, and linear-causal paradigm of classical science" (p. 421). Most refer to the phenomenon of systems theory as understanding that the "whole is more than just the sum of its parts," which is the additive interpretation of Koffka's (1963) principle.

Contemporary uses of systems thinking have been fairly broad. To date, application of systems theory in our field has helped us examine the interactional relationships for influence among the social actors, norms, routines, and structures within specified temporal and spatial contexts. For instance, systems theory has been applied to studies on state policies, district reform, school improvement efforts, and leadership practices. Studies focused on each of these levels, or subsystems within education, have provided important insights into how organizational and institutional change does or does not happen within specified boundaries (Daly & Finnigan, 2016; Rorrer, Skrla, & Scheurich, 2008).

However, by focusing on a subsystem (e.g., solely on the federal, state, school or district levels) in the absence of the total system, our field has maintained a void in understanding the larger interactional relationships for influence, interdependence, and dynamics, as well as the power of these subsystems working together, within and among the subsystems themselves. As each level of the system gets treated as a distinct player rather than in the system, the larger holistic organism remains neglected. We wonder if this tradition has led to difficulties in reform implementation and the transformational abilities of a reform. Thus, although we value the earlier work that has applied systems theory, we argue that through a dynamical systems theory (DST) approach we actually get closer to understanding what Gestalt psychologies Kurt Koffka (1963) intended with the idea behind "the whole is other than the sum of its parts" (p. 176). In this regard, the "whole" attains an entirely separate being from the parts. In utilizing DST, we avoid an explanation of change that is devoid of all the rudiments and complexities that converged to create the issue, or problem, in the first place and the complexity in resolving or bringing about change.

Consistent with our reference to the importance of Koffka's (1963) expectation for transformational change, Morgeson and Hofmann (1999) further prompt us to redirect our attention to questioning what aspect of the system we focus upon and consider a broader entity as the system. We find their explanation of the role of the collective as a compelling argument for focus on the group involved as a whole. They explain, "as interaction occurs within larger groups of individuals, a structure of collective action emerges that transcends the individuals who constitute the collective" (p. 252). We believe this transcendence is perhaps among the most critical aspect of achieving the potential for turnaround.

Explicitly, it is the very nature of possible collective action and interdependence within the system that compels our exploration of what are the relationships that may help us achieve turnaround that is scalable and sustainable. Thus, having previously identified the leadership moves that support reclaiming turnaround (Park et al., forthcoming), we now pose the question: what conditions allow for momentum and collective action within the system to achieve turnaround? As we began to examine the turnaround process in our state, it became apparent that in addition to an emerging alignment in the expectations for turnaround, we noted that all subsystems were moving simultaneously, although not always in the same direction. Though this movement isn't always coherent across the system, the existence of collective action throughout the system opened new opportunities for substantive change and at times conflict. Part of the emerging alignment in the expectations for turnaround that we observe is a centrality of acknowledgement for turnaround. This does not necessarily mean "acceptance" of turnaround nor "readiness" for turnaround. Instead, we have observed a mutual nod that improvement is necessary. As we explain further, while there remains divergence in what counts as improvement, or turnaround, there are no longer factions within the polity that deny that it needs to occur.

In order to extend the application of systems theory to turnaround, we situate our analysis in the underpinnings of DST, or what is commonly known as chaos theory. As noted within the theory, chaos is not indicative of the absence of order. Instead, a dynamical system is, as Grobman (2005) explains, "a system whose state at any given point in time determines its state at a future point in time and does not reach either a fixed point or an equilibrium" (p. 365). DST moves our explanations of change in school turnaround, from linear, causal explanations to one that explores the patterned sequence of actions that occur and the results of the patterned sequence of events (Abbott, 1990).

## *Non-linearity and Variable Coupling within DST*

Traditional models of change are founded on knowing, or being able to surmise, the beginning point of change as a static point in time. Furthermore, many models of change ascribe to explanations of change that occur in somewhat predictable and linear ways, or "phases" (e.g., Lewin, 1947). Yet, Morel and Ramanujam (1999) indicated that it was more common that organizations were "dynamic systems of adaptation and evolution that contain multiple parts which interact with one another and the environment" (p. 278). Despite this, there continues to be an over-reliance on linear explanations and expectations for change in education.

In turnaround we observe the principle of nonlinearity, which is more compatible with the complexity and interrelatedness of DST (Casti, 1994; Daft & Weick, 1984; Lewin & Volberda, 1999; Morel & Ramanujam, 1999; Piehl, 1974). DST indicates that organizational changes occur in less than predictable ways, which also disrupts the notion of the linear phase explanations of turnaround. Instead, as DST contends, change occurs through observable patterns that emerge from a series of small, successive changes. As indicated by Dooley and Van De Ven (1999), "When this occurs, the resulting behavior has come to be called 'chaos,' to distinguish it from truly random behavior" (p. 360).

Linear explanations may be easier to understand, particularly through a retrospective telling of how change occurred. However, we contend that transformations in educational organizations emerge through nonlinear, discontinuous relationships, networks, and actions, and ultimately reflect more complex and intricately woven aspects of organizational life. This quality requires attention to the continuous adaptation of actions that occur with shifting internal and external conditions. Moreover, the flow of information, demands, actions, actors and their relationships, and their patterns create new possibilities for disruption and transformation. While this doesn't lead to seemingly predictive next steps, understanding the messiness of turnaround through this DST lens provides an opportunity for creating stability that is consistent with the intended outcomes.

In addition to the requisite expectation for non-linearity in turnaround, we add the need for attention to variable coupling within the system for turnaround to occur. Many have relied on Weick's (1976) early depiction of education as a loosely coupled system, where he highlighted the agility of loosely coupled systems to respond to their environment. Although not often referenced in the conversation of education as a loosely coupled system, Weick also invited us to consider a point relevant to our use of DST and nonlinearity in explaining turnaround. He noted "a tight coupling in one part of the system can occur only if there is loose coupling in another part of the system" and that "it may be the pattern of couplings that produces the observed outcomes" (p. 10). With this in mind, we highlight reasons why variable coupling within the system is advantageous. Diversity in loose and tight couplings ensures a degree of flexibility in the ability of the system to respond, or adapt, as actions occur and environmental conditions shift (Lotto, 1983). Next, variable coupling also ensures that both the individuals and the collective can maximize their engagement, effectiveness, and efficacy in the improvement efforts. Third, variable coupling can be strategically determined and thereby used to achieve outcomes.

## Methods

This chapter is drawn from a study using participant observation methods (DeWalt & DeWalt, 2011). Participant observation is used frequently in anthropological, ethnographic and organizational studies. Using turnaround as the case, we explored the turnaround phenomenon in Utah from an insider and outsider perspective. Here we have drawn three phases, as noted by Stocking (1983), which are key to participant observation studies: participation, observation, and interrogation.

We are overt in our participant observation methods, as all participants and entities are notified as part of our work that this may be included in research. As participants, we have served as facilitators of school turnaround support at multiple sites and the state-wide turnaround leadership development program, known as Leadership for Inquiry and Turnaround (LIFT). As researchers, we have served as the evaluators for a history of state education policy, which included the school improvement and turnaround policies.

We explored the changes in school improvement and turnaround initiatives in the state, the various actors engaged in these initiatives, and the outcomes of such efforts using self-reported exit tickets/surveys, document and policy archival data, interview data, notes from planning meetings, and school support session notes collected between 2015 and 2018. The analysis shared here is a reflection of the lessons we have learned as both participants and researchers through both engaging in and studying turnaround reform efforts and the interrogation of the corpus of data relative to "what conditions allow for momentum and collective action within the system to achieve turnaround?"

## The Case of Turnaround

Set within the national policy context, this chapter focuses on the turnaround efforts in Utah. In addition to the federal SIG stipulations that were implemented in state, district and school improvement plans, school turnaround in Utah received more attention when Utah's legislature passed SB235 in the 2015 Utah Legislative Session. SB235 stipulated that schools performing in the bottom three percent within Utah's accountability system would be designated as needing turnaround and given up to three years to improve student performance. (The first cohort of 26 "Turnaround Schools" was identified in fall of 2015, and includes both charters and traditional schools. An additional 17 were identified in 2016. Many of these schools were previously or simultaneously identified as Title I Priority or Focus Schools.) SB235 also included requirements for schools to establish a school turnaround committee, engage in root-cause analysis, and made stipulations for "failing to improve school performance" (School Turnaround and Leadership Development Act, SB235, 2015). Similar to the federal requirements previously established, failure to achieve increased student academic performance could lead to further high stakes sanctions (e.g., restructuring through conversion to a charter, contract management group or state takeover, termination of a charter, closing a charter, transferring operations to another school). SB235 also provided for school recognition and a rewards program that included additional financial incentives for increased student performance. An additional requirement embedded in the state legislation included the need for schools identified for turnaround to contract with an external turnaround specialist who was required to assist them with the development and implementation of the turnaround plan.

## *Interactional Relationships and Influence in School Turnaround*

Through both our research of turnaround and our work as an adaptive systems partner supporting turnaround, we have observed four types of interactional relationships for influence that exist within the system – confluence, collision, push, and pull. These types of relationships are not just about the actors or between subsystems, but also the relationships between tasks. In part, we have observed that the coexistence of these types of interactional relationships contribute to the iterative changes necessary for school improvement to occur. These relationships may either assist or hinder progress and transformation.

Among these four types of interactional relationships for influence, we have identified three key dimensions for each: (1) directionality of influence and change: (2) the power and reciprocal responsibility, degree of reciprocity and power; and (3) primary interactional style. Along these dimensions, we propose a typology of systems relationships that results (see Table 11.1: Typology of Systems Relationships). The directionality of influence and change highlights the degree of shared purpose, goals, and even actions. The power and reciprocal responsibility highlights the degree of mutual interdependence, power, and influence. And the primary interactional style highlights the main norms and nature of the relationship. The types of interactional relationships and their dimensions are described below.

The first type of interactional systems relationship is confluence. When we refer to confluence, we mean that all elements of the system are moving in the same direction, towards the same goals. Furthermore, there is a high level of reciprocity and interdependency within the subsystems. Power and decision-making authority are shared. The interactional relationships for influence between the subsystems are primarily characterized by co-construction.

The second type of interactional systems relationship is *collision*, where parts of the systems are moving towards one another, but at cross-purposes. Rather than mutual dependence and reciprocity, power and influence is fragmented. This leads to an interaction style that is primarily characterized by conflict.

The third type of interactional systems relationship, which we refer to as *pull*, occurs when one part of the system drags the other parts towards a shared goal. Instead of mutual reciprocity, power is unevenly distributed, with one element of the system having greater influence. Responsibility for creating change is also unevenly distributed, typically with one subsystem wielding more power while another subsystem has greater responsibility with less power. Thus, the main interactional mode among the part of the system is resistance, reactive, or compliance-oriented.

Finally, we refer to the fourth type of interactional systems relationship as *push*. Similar to the pull interactional systems relationship, one part of the system is propelling the other parts towards a desired goal and reflects an unbalanced distribution of power and reciprocity in responsibilities. The interactional relationships between the parts of the system are primarily coercive or characterized as resistant.

By no means is the typology of interactional relationships and dimensions meant to be exhaustive, or mutually exclusive. They are not binaries. These relationships are fluid. For example, we have found many instances in which multiple types of interactional relationships for influence occur simultaneously as the system moves. That is, the existence of one type of interactional relationship does not preclude the existence of another within the same system at the same time. Moreover, state, district, and school responses vary, and vary over time, actions, and actors. Depending on context, demands, and expectations, performance and capacity may vary. For instance, two turnaround schools in the same district may have vastly different staffing needs – with one experiencing high teacher turnover and another experiencing a stable but aging teaching staff. This in turn may influence the types of relationships between districts and schools (e.g., one may be more collision vs. pull).

In the following section, we explore the application of DST and the typology of interactional relationships by describing five high leverage areas of turnaround that we have observed in Utah. The five areas include: establishing priorities, strengthening leadership, tuning professional learning, utilizing improvement cycles, and maximizing relationships. Within each of these leverage areas, we provide examples of where we have observed confluence, but also examples of other interactional relationships. Taken together, these five high leverage areas illustrate how the system is engaging with turnaround. The efforts within these five areas are not linear, nor are they strategically planned holistically. We have also observed different types of interactional relationships and various degrees of coupling between leverage areas as well as between the subsystems within and among the leverage areas.

**TABLE 11.1** Typology of Systems Relationships

| *Type of Interactional Relationship for Influence* | *Dimensions* | | |
|---|---|---|---|
| | *Directionality of Influence and Change*<br>*Degree of shared purpose/goals* | *Power & Reciprocal Responsibility*<br>*Degree of mutual interdependence, power, influence* | *Primary Interactional Style*<br>*Main Norms and Nature of Relationship* |
| *Confluence* | All parts of the system moving towards same goal through iterative interactions | Mutual interdependence and influence between parts of the system, shared power and decision making | Co-constructive, iterative |
| *Collison* | Parts of the system are moving towards each other but at cross-purposes | Fragmented power/influence; power struggles; minimal interdependence | Conflictive, transactional |
| *Pull* | On or more parts of the system are dragging the other towards a goal | Uneven power/influence – one part of the system is driving/propelling the other parts; uneven responsibility | Compliance-oriented, resistive or reactive |
| *Push* | One or more parts of the system are propelling the other(s) towards a goal | Uneven power/influence – one part of the system is forcing or prodding | Coercive, resistive by one part of the system |

## *Establishing Priorities*

Despite a history of high expectations for student outcomes within the state, there has also been a rationale for when the performance of all students was not consistently high that focused on the "challenging" contexts as limiting factors for school improvement efforts. With new calls for school turnaround, the ambiguity of expectations for student outcomes across student groups has evolved into more taut expectations for explicit targets within the area of teaching and learning. Throughout the growing school turnaround efforts in Utah, we have noticed the intentional focus on establishing such priorities.

There remains a recognition that efforts across multiple dimensions must occur simultaneously for turnaround to be successful. As each of the subsystem levels have taken their own loosely coupled approach to engaging in improvement efforts, turnaround has demanded that there be established expectations and processes for deciding and enacting priorities. In part, this focus has been evident through expectations for explicit student achievement targets, shifts in language and nature of communications about the potential capacity for student and school success, and delineation of high priority actions that, if completed, will result in significant changes.

Among the shifts that have occurred are the fore-fronting of a student-focused approach that relies upon a tightly coupled system of support from teachers, coaches, and principals. We have observed these shifts illustrated by principals who initiate messages to their whole school staff, such as "What unifies us is our commitment to supporting all students' growth." Additional evidence of this shift is in the language about standards-based teaching from the question, "What am I teaching?" to "What are students learning?" In part, this shift has occurred through efforts to provide intentional messaging, clarify role definition and expectations for enactment, and maintain monthly implementation cycles that culminate in shared data gathering and analysis.

Establishing priorities with turnaround has occurred also through the use of data for identifying instructional strategies and processes to achieve planned outcomes (Datnow & Park, 2009). Within turnaround in Utah, this occurs through the state adopted method of short, rapid cycles for improvement. The 90-day cycle for improvement has become ingrained in the nomenclature and the behaviors of educators working in low-performing schools. In addition to the 90-day planning cycles (and sometimes as part of 90-day plans), we have observed the prioritization of time for educators to plan, study, and analyze data to track students' progress on mastery of core standards. These priorities translate to classroom practice in the form of instructional strategies, processes, and routines, for example, schools who take on the implementation of one or two evidence-based instructional strategies (e.g., student opportunities to respond and formative assessment through daily checking for understanding) as a whole faculty by learning about the strategies and then monitoring how the strategies are being implemented and resulting in student learning.

Priorities have also been established by teams of teachers, principals, and instructional coaches working together in collaboration to identify each person's actions to support student achievement, including how teachers and coaches will collaborate around data gathering and analysis. School districts have initiated deliberate structures for teacher collaboration to enhance instruction (e.g., Professional Learning Communities). Collaborative structures, such as weekly PLC meetings, have included attention to group norms and identifying individual responsibilities while developing shared responsibility among group members. Within this structure are increasing expectations for teachers to use their collaboration time to focus on powerful student learning opportunities, which are often influenced by district protocols and guidance for how teachers use their time and what they focus on during their meetings. For example, we have observed a focus across districts on the importance of shared intentional planning for first (Tier 1) instruction using planning processes and content standards, beginning with the question "What should students know and be able to do by the end of the lesson?" The coupling between the districts and the schools here varies, as some districts are more prescriptive than others.

With the focus across districts on teacher collaboration to support instruction, tensions have emerged related to the ways teachers have been supported to implement district-mandated structures and protocols, as well as the interpretation of district-wide initiatives that sometimes compete with the teacher collaboration structures developed by the schools and their school improvement plans. For example, one district central office created an extensive curriculum map of over 200 pages per grade level that identified eight evidence-based instructional priorities (e.g., explicit instruction, teacher clarity, structured classroom discussions, etc.). District expectations for implementation of the eight instructional priorities were messaged during a whole district "roll-out," where school personnel received a message pushed by administration during a district-wide professional development day. The message from the district was, "It is non-negotiable that all teachers adhere to the instructional priorities and implement them in daily instruction. Principals – you are expected to monitor implementation through weekly Instructional Priorities Observations, and provide teachers feedback." During conversations at four schools with teachers in their collaborative meetings and job-embedded professional learning designs, teachers' awareness and knowledge of the eight instructional priorities was limited, as evidenced by not knowing the location of the instructional priorities in the curriculum map, or how to implement during classroom instruction. Principal observation data also revealed that the eight instructional priorities were not being implemented in a way that resulted in increased student performance.

As instructional coaching has grown as a priority, so have the intentionality of written and verbal communication about the purpose of coaching, how coaching is enacted, and the supports (e.g., professional learning opportunities) for coaches. All districts have implemented a variety of coaching models, often targeted in Title I schools. As a result, influences on social norms and culture and climate have also occurred. There have been many tensions as schools design independent coaching systems and structures that may or may not be consistent with district expectations about the role or purpose of coaching. This tension has been especially prevalent in the role of coaching and teachers' coverage of the curriculum versus students' mastery of core standards. For example, the UEPC approach to coaching draws from recent research and best practices in instructional coaching which emphasizes establishing a shared, understood, enacted, and evaluated model by all members of the system. For some districts and schools, the coaching models lack a unified approach with no evaluation system to understand the impact of coaching on student learning. Thus, in some cases, there are different coaching models enacted not just between UEPC and districts but also between districts and their schools. These disconnects illustrate the tensions between priorities set at the school level separate from the district levels.

## *Strengthening Leadership*

The mantra, and more importantly the research, that "Leadership Matters" has become pervasive in discussions and support of school turnaround. There has been confluence within the system to acknowledge that while teachers may have the most significant direct impact on student achievement, effective leaders are necessary to create the conditions for instructional success and improved student learning. This is an important move that has led to the need for distributed leadership models supported by developing core leaders throughout the schools. Building capacity in this way has also been used to identify and improve commitments to priorities for teaching and learning.

Since the turnaround initiatives in the state began in full force, there has been a concerted effort throughout the system to strengthen leadership. These efforts have targeted multiple points of development for leadership, including from identifying aspiring principals to focusing on university-based preparation programs to development of leaders in practice. There are a number of loosely coupled initiatives in motion, such as aspiring leadership academies, professional learning for leaders in practice, professional learning and support for teacher leaders and coaches, and even use of external partners to support the needs of leaders in turnaround more specifically.

In addition to the state and district level approaches to improving leadership, the UEPC provides a network improvement approach to leadership development. The UEPC LIFT (Leadership and Inquiry for Turnaround) is part of the statewide leadership capacity building efforts of schools who have been identified as "turnaround schools" (i.e., performing among the lowest 3% of schools in the state). LIFT emphasizes amplifying leadership in four domains: turnaround leadership, instructional transformation, culture shift, and talent development (Center on School Turnaround, 2017). To achieve these areas of growth and effectiveness, LIFT builds five leadership capacities as a means to sustainable improvement: clarity of focus and unifying purpose, inquiry and reflection, relationships and culture, processes and structure, and communication (UEPC, 2018). Recognizing the need to encompass a more holistic approach to leadership, LIFT moved from a model focused on the principals to the school leadership team. LIFT, which provides ongoing professional learning and action planning to enhance turnaround efforts is purposefully tightly coupled to the other high leverage areas outlined here, and is enacted in a way that brings all parts of the system together. In an effort to create opportunities for confluence, the UEPC has worked with the state office of education personnel and included district and school leaders, and policymakers in the planning and participation of LIFT.

While strengthening leadership has been emphasized at all levels of the system, these efforts to support leadership development have not always been intentionally aligned or coordinated. As such, schools often experience tensions when they are involved in multiple leadership initiatives that might prioritize different emphases or use different language related to improvement strategies or processes. For example, schools who participate in the UEPC's LIFT program also work in school districts and with turnaround partners who offer different types of guidance about turnaround that may not be consistent with the messages schools hear in the LIFT program or other district- or state-sponsored leadership initiatives. Attempts to increase alignment and coherence, such as our invitation to district representatives and turnaround partners to participate in the LIFT program, have had varying degrees of success in fostering alignment and coherence with different districts and turnaround partners, particularly with those who tend to be grounded in the same research and frameworks and others who prescribe. While disconnects are still observed, there is momentum now within the turnaround efforts across the system to be purposeful in communications and planning, such as efforts to cross-walk between initiatives and intentions, to avoid contradictions and resistance that might undermine optimal outcomes.

## Tuning Professional Learning

While the importance of professional learning has remained a high leverage area for turnaround, state and local resources for professional learning, unfortunately, have diminished over time. As competition for funding continues, schools in turnaround are increasingly tuning their professional learning to align with identified priorities and goals related to teaching and learning. Specifically, we are witnessing several shifts begin.

First, professional learning is less a string of random events and instead is based on the needs of teachers to meet the needs of students. Next, job-embedded professional learning designs are being supported. These experiences are created for teachers to increase knowledge, skills, and dispositions for increasing student clarity of lesson objectives and success criteria, and close the gap in student performance, particularly for English learners, students with disabilities, and other historically under-performing student groups. Job-embedded professional learning, which is gaining momentum, is often less expensive, situated in everyday practice, authentically connects teacher practice to student impact, and actively involves teachers, coaches, and principals in shared, inquiry-based and solution-oriented learning, resulting in the ability to diagnose and respond to student needs.

In addition, teachers are gaining choices in the professional learning opportunities available to them, including professional learning offered by peers, district and state offices of support, and external partners. Moving from compliance to agency, several factors influence teacher decisions about learning designs, including the teachers' learning style, the extent of the intended change for students, the school context, and resources available to support learning.

The implementation of authentic professional learning communities based on inquiry, review of instructional and student data, and aimed at supporting professional dialogue centered on teaching and learning, also offers educators the opportunity to develop learning. One of the primary challenges with PLCs has been the shift from a gathering of teachers to a meaningful structure where professionals are truly learning in community. The ideal characteristics of a group of professionals learning in community include shared mission, values, and goals, a focus on learning, collective inquiry, results-oriented progress, and a commitment to continuous improvement. One of the primary tensions that exists here is the contrast between these ideal characteristics and the current state of collaborative structures in turnaround schools where these ideal characteristics do not exist. In part, we find that several interactional relationships impeded these ideal characteristics, including the existence of a push for mandated professional learning, a collision between the state, district, and school over limited resources such as time and money, and a pull by districts and school over who would offer specific professional learning.

Another key example of the tensions that we have observed around the approach to tuning professional learning, especially in the form of job-embedded structures such as PLCs, is the focus on standards-based instruction among turnaround schools. As noted above, across multiple levels of the system, turnaround schools have been focusing on common instructional strategies that have included a focus on supporting students' mastery of core standards. This has been a focus on individual school improvement plans and is promoted by district, state, and turnaround partners. We have observed particular tension in the design of professional learning opportunities for teachers that often pits the importance of "coverage of learning content" against the importance of "student mastery of standards." The UEPC approach to coverage versus student mastery draws from national research on effective practices in urban schools (e.g., see Johnson, Uline, & Perez, 2014) which suggests schools who are able to get all groups of students to achieve at high levels focus on ensuring students master standards in depth rather than simply covering content. In this approach, teachers' collaborative planning of lessons is critical to ensure that lessons are designed to support student mastery during first instruction. Some districts are still grappling with focusing on mastery of key standards and often the districts' message is that teachers follow the curriculum map and stay on pace to "cover" all standards by the state testing window. We have yet to see unification of, or confluence of, these approaches, and consequently tensions exist for school-based educators who hear different messages.

## *Utilizing Improvement Cycles*

As mentioned previously in the establishing priorities section, utilizing data and improvement cycles has become a high leverage area and an area of confluence in turnaround efforts. Although working through improvement cycles per se is not new in the state, 90-day rapid improvement cycles are becoming more usual, particularly cycles intended to improve instruction and learning. The 90-day cycles generally include three one-month implementation cycles with six stages. The stages reflect the following: 1) investigate and use data to identify intended student changes, 2) plan for targeted professional learning for teachers to build capacity to realize student changes, 3) allow time for teachers to practice application of new strategy, process or routine with support (coaching, peer coaching, ongoing professional learning), 4) assessment of implementation, 5) reflection on student outcomes, 6) modification and revision of implementation for the subsequent month or 90-day cycle. The State Office of Education provided the template for the 90-day plan based on their work with the University of Virginia Turnaround Program (Player, Kight, & Robinson, 2014).

Although there is a foundation, even strong expectations, for data-driven decision making, creating and maintaining a culture of data use through the use of formative and summative data is a work in progress and one where tensions exist. The state has significant infrastructure invested in information management systems. This system makes standardized data available to all schools annually. In addition, the state has offered "Data Driven Institutes" to support the lowest-performing schools in using data-driven instructional practices that include feedback and observation cycles and use of formative or quarterly benchmark assessment data that are available through the various assessment platforms. Districts have varying degrees of infrastructure, and often infrastructure is based on available resources. Currently, districts have invested in a variety of student information platforms that allow teachers to access ongoing benchmark and state assessment data to use in monitoring student progress and designing interventions as needed. However, the time available and skill to utilize these data to their maximum depth vary greatly.

Despite these investments in technology and management systems, the data available from these systems is different from the types of data necessary at the school and classroom level to make decisions for improvement in rapid cycles. Currently, the use of interim or benchmark data is often loosely coupled to planning for instruction. This results in a loose coupling between standards-based instruction and assessment. Another repercussion of this loose coupling is an often haphazard approach to diagnosing and responding to students' needs using reliable and timely data sources. We note that a confluence by school, district, and state levels about each other's expected roles in providing timely, accessible data and building capacity for data-driven decision-making to engage in continuous improvement that permits teachers to make instructional decisions based on student needs remains unattained.

### *Maximizing Relationships*

A final area of leverage that we have observed in turnaround is the work to maximize relationships. In addition to expanding the type of relationships that exist within and across the system, we note an important shift in the nature of those relationships. For instance, these relationships are becoming more collaborative and generative in nature as compared to adversarial or territorial.

An important aspect of the relationships among the "players" in the turnaround system throughout the state is that they have been established systematically. For instance, the UEPC as an external and adaptive assistance partner has formed relationships with teachers, leaders, schools, districts, and the state education officers and policymakers. These professional relationships are seen as necessary and are engaged in as an interdependency within the system. In part, these relationships have been formed by a confluence around intended changes and improvements for students in schools. Our relationships are not based on compliance in a hierarchical system, and are designed to create systemic coherence between all the entities to eliminate barriers, and to provide pathways to support changes in turnaround schools. In doing so, networks are formed and provide an opportunity for important insights into the practices and strategies that are utilized by the most successful schools. Through the network of educators and leaders involved in turnaround efforts throughout the state, we find that strategies and processes for cultivating professional relationships are named, and explicitly refined through protocols, written messaging, and communication tools.

The LIFT program that we host provides an example of how a network of educators and leaders who regularly engage around turnaround has maximized relationships. Although initially schools entered LIFT through push and pull, through the regular interactions at the LIFT meetings, as well as through the electronic communication and resources provided through the program, we have observed an increase in the degree to which members of school teams have reached out to each other to share resources, request help with challenges, and create opportunities for connecting and collaborating across district boundaries and even across remote areas of the state. For example, one of the principals involved in LIFT developed a turnaround coaching PLC based on ideas that were generated by her team at one of the LIFT meetings. She sent out invitations to other principals in the LIFT network about the opportunity for instructional coaches at other schools to participate in a monthly (virtual) turnaround PLC. This relationship has resulted in monthly collaboration between instructional coaches from the northern and southern parts of the state who would have not come together otherwise.

As we have observed relationships become more collaborative and generative in nature as compared to adversarial or territorial across the system, we have continued to observe tensions in the ability to maximize relationships due to lack of structures or opportunities for regular and coordinated communication that is specifically designed for increasing alignment and coherence through the system. A logical entity to take on this coordination and communication role might be the state office. The state office of education provides regular communication about updates on state and federal turnaround policies, engages in ongoing monitoring and compliance of programs, and developing tools and resources for enacting the required state turnaround procedures. However, with smaller relative state fiscal investments in administrative support, their capacity to engage school, district, and turnaround partners throughout the state to maximize relationships is constrained. As such, informal networks remain the key mechanism connecting turnaround actors in the system, which does not tap the potential for a more coherent and aligned system.

## Conclusion

Neither the federal nor state policies for turnaround *required* a systems approach nor stipulated specific roles for subsystem actors. These omissions are an obvious void, particularly given the possibilities for transformation when all aspects of the system – the subsystems – work collectively to achieve improved teaching and learning rather than relying on any one subsystem to carry the responsibility for turnaround. In addition, this void in policy implicitly indicated that turnaround must be, or should be, top-down rather than bottom-up or a combination of top/down and bottom/up.

Initially, we observed turnaround efforts in the state steeped in compliance, conflict, and resistance, which limited opportunities for "turnaround as transformation" or as an effort that prompted substantive, deep-level changes meant to alter the experiences of teachers and students. Our aim here, and our work in turnaround generally, was to engage in an inquiry that informs a more nuanced understanding of how turnaround occurs and what factors within the system influence the possibilities for progress and change.

In part, this inquiry is aided by consideration of turnaround from a systems approach. The strength of systems theory, generally, as an analytical framework in education is that it continues to push us from consideration of, for instance, school improvement/turnaround as discrete events, episodes, or actions rather than as contextually embedded and connected elements. Despite the presence of systems theory in our field, we still witness the quest to over-simplify the process of change and improvement to something that can be a linear and abridged process that will produce the "one best solution." School turnaround is no exception.

Yet, an understanding and application of dynamical systems theory in particular helps us avoid the trappings of previous explanations of change, specifically the all too simplistic, normative and prescriptive recommendations for turnaround. Dynamical systems theory reminds us that change is part of the natural order of being for an organization. Moreover, using the principles of dynamical systems theory as an analytical framework to consider turnaround reform and its potential, we have discovered the demand for the system to be both predictable and responsive while also being flexible and adaptive across actors, structures, functions, and procedures. Furthermore, our exploration indicates that the varied interactional relationships between and among the subsystems, particularly within high leverage areas, demand attention to the nonlinearity, interdependencies, and variable coupling among efforts and the subsystems. Recognizing the interdependence within the system and that the subsystems (e.g., school) cannot gain momentum on improvement efforts without reciprocal changes in other subsystems (e.g., district, state, national) requires a different focus. Taken together, these interactional relationships establish opportunities for *influence and dynamic disequilibrium* within turnaround.

One lesson learned from school turnaround is that the system may experience aspects of all four interactional relationships for influence simultaneously. Interestingly, we have observed that each type of interactional relationship can serve a greater utility in moving the system towards confluence, which seems to be when more optimal levels of change occur. The co-existence of the various interactional relationships serves as a means to clarifying values, building cohesion and collective identity, norming organizational members, prompting necessary changes, and even extinguishing resistance.

That said, of the four types of interactional relationships, we are particularly intrigued by the possibilities for a *confluence of influence* within the system. The *confluence of influence* interactional relationship occurs when the subsystems are moving towards the same goal through on-going interactions, when there is interdependence within the system and power and decision-making are shared. We observe that confluence of influence most often occurs when the system has agility to learn and adjust to meet the needs of all subsystems, refine its ability to be both goal oriented and improvement-centered, and maintain momentum through agile couplings between system levels, areas of leverage, and actions. One limitation of our work to date is understanding whether or not the antecedents to confluence can be strategically designed prior to implementation or whether more organic emergence of factors is necessary.

This inquiry, and our work with turnaround more generally, indicates that we have to stop promoting the idea that improvement is about reaching the next plateau. The ideal of equilibrium and balance is misleading. Education is in constant battle with the institutional forces that promote either conditions for "business as usual" or that advocate for massive change. We recognize that this inquiry will not extinguish the appetite to create a hierarchy of importance in understanding the roles of system levels or promoting their imperative existence in school turnaround efforts. Since any momentum gained in turnaround schools is tenuous, further inquiry into the nature of influence will be necessary. Research is also needed to better understand how a system can avoid decline or negative redundancy in the interrelationships and system efforts, and how changes among the relationships may alter efforts to achieve turnaround.

We extend further to say that perhaps the optimal goal for turnaround is recognizing the promise of a dynamic disequilibrium. The state of dynamic disequilibrium occurs because of both the instability in the internal as well as external environment, including the nature of the interactional relationships. By casting system change in the context of dynamic disequilibrium, we can finally embrace and optimize the principles of nonlinearity, variable coupling, and interdependence as strategic conditions. The possibilities created by embracing a dynamic disequilibrium that experiences iterative cycles of a confluence of influence include the potential for collective action and ensuring a "whole [that] is other than the sum of its parts." We recognize that promoting this may be a limitation for field application, particularly given the desire among practitioners and policy makers alike for linear implementation models with expectations for rigid implementation fidelity in improvement and school turnaround.

For far too long, scholars (and practitioners) have sought simplistic understandings, and subsequent renderings of what was the right mix of events, people, and actions that created the transformation they are retrospectively analyzing. Instead, we offer that there remain important questions about the nature of the inter/intra relationships within the system and among actors in the system, the actions that occur, and the values, beliefs, and norms that are established, and the triggers for the system to become imbalanced and then self-correct. Finally, we are compelled further to ask questions that help us understand how to demonstrate that system complexity is not a shortcoming, but rather a strength that we can acknowledge and harness in turnaround settings.

## References


Abbott, A. (1990). A primer on sequence methods. *Organization Science, 1*(4), 375–392.

American Recovery and Reinvestment Act (ARRA) of 2009, Pub. L. No. 111-5 § 123 Stat. 111, 115, 516 (2009).

Bertalanffy, L. von (1972). The history and status of general systems theory. *The Academy of Management Journal, 15*(4), 407–426.

Casti, J. L. (1994). *Complexification: Explaining a paradoxical world through the science of surprise.* New York: Harper Collins.

The Center on School Turnaround. (2017). *Four domains for rapid school improvement: A systems framework.* San Francisco, CA: WestEd.

Daft, R. L., & Weick, K. E. (1984). Toward a model of organizations as interpretation systems. *Academy of Management Review, 9*(2), 284–295.

Daly, A., & Finnigan, K. (2016). *Thinking and acting systemically: Improving school districts under pressure.* Washington, DC: American Educational Research Association.

Datnow, A., & Park, V. (2009). Conceptualizing policy implementation: Large-scale reform in an era of complexity. In G. Sykes, B. Schneider, & D. Plank (Eds.), *AERA handbook of education policy research* (pp. 348–361). New York: Routledge Publishers.

DeWalt, K., & DeWalt, B. R. (2011). *Participant observation: A guide for fieldworkers*. Lanham, MD: Rowman and Littlefield.

Dooley, K., & Van de Ven, A. (1999). Explaining complex organizational dynamics. *Organization Science, 10*(3), 358–372.

Dragoset, L., Thomas, J., Herrmann, M., Deke, J., James-Burdumy, S., Graczewski, C., Boyle, A., Upton, R., Tanenbaum, C., & Giffin, J. (2017). *School improvement grants: Implementation and effectiveness. NCEE 2017–4013*. Washington, DC: National Center for Education Evaluation and Regional Assistance.

Elementary and Secondary Act of 1965, Pub. L. No. 89–10 (1965).

Every Student Succeeds Act of 2015, Pub. L. No. 114-95 § 114 Stat. 1117 (2015–2016).

Grobman, G. (2005). Complexity theory: A new way to look at organizational change. *Public Administration Quarterly, 29*(3/4), 350–382.

Johnson, J. F., Uline, C. L., & Perez, L. G. (2014). The quest for mastery. *Educational Leadership, 72*(2), 48–53.

Koffka, K. (1963). *Principles of Gestalt psychology*. New York: Harcourt, Brace & World.

Lewin, K. (1947). Group decision and social change. *Readings in Social Psychology, 3*(1), 197–211.

Lewin, A. Y., & Volberda, H. W. (1999). Prolegomena on coevolution: A framework for research on strategy and new organizational forms. *Organization Science, 10*(5), 519–534.

Lotto, L. S. (1983). Revisiting the role of organizational effectiveness in educational evaluation. *Educational Evaluation and Policy Analysis, 5*(3), 367–378.

Morel, B., & Ramanujam, R. (1999). Through the looking glass of complexity: The dynamics of organizations as adaptive and evolving systems. *Organization Science, 10*(3), 278–293.

Morgeson, F., & Hofmann, D. (1999). The structure and function of collective constructs: Implications for multilevel research and theory development. *Academy of Management Review, 24*(2), 249–265.

No Child Left Behind Act of 2001, P.L. 107-110, 20 U.S.C. § 6319 (2002).

Park, V., Groth, C., Bradley, J., & Rorrer, A. (forthcoming). Reclaiming turnaround for democratic schooling: Leadership moves to build capacity for teaching and learning. In C. Meyers & M. Darwin (Eds.), *International perspectives on leading low-performing schools*. Charlotte, NC: Information Age Publishing.

Piehl, D. (1974). Nonlinearity and the study of organizations. *Academy of Management Journal, 17*(4), 760–767.

Player, D., Kight, M., & Robinson, W. (2014). The state's role in supporting data use to drive school turnaround. In L. M. Rhim & S. Redding (Eds.), *The state role in school turnaround: Emerging best practices* (pp. 101–113). San Francisco, CA: WestEd.

Plsek, P. E., & Greenhalgh, T. (2001). Complexity science: The challenge of complexity in health care. *BMJ: British Medical Journal, 323*(7313), 625.

Rorrer, A. K., Skrla, L., & Scheurich, J. J. (2008). Districts as institutional actors in educational reform. *Educational Administration Quarterly, 44*(3), 307–357.

School Turnaround and Leadership Development Act, Teacher Quality Amendments, HB 328, § 53A-17A-163 (2015).

Shaked, H., & Schechter, C. (2013). Seeing wholes: The concept of systems thinking and its implementation in school leadership. *International Review of Education, 59*(6), 771–791.

Stocking, G. (1983). The ethnographer's magic: Fieldwork in British anthropology from Tylor to Malinowski. In G. Stocking (Ed.), *Observers observed: Essays on ethnographic fieldwork* (pp. 70–120). Madison, WI: University of Wisconsin Press.

U.S. Department of Education (2009). *Race to the Top Executive Summary*. Washington, D.C. Retrieved from https://www2.ed.gov/programs/racetothetop/executive-summary.pdf

Utah Education Policy Center (2018). LIFT. Retrieved from https://uepc.utah.edu/lift/

Weick, K. (1976). Educational organizations as loosely coupled systems. *Administrative Science Quarterly*, *21*(1), 1–19.

# 12

# IMPROVING SCHOOLS IN VICTORIA, AUSTRALIA

## System, Region, and School Perspectives

*David Gurr and Lawrie Drysdale*

The stimulus for this chapter was an earlier paper in which we were charged with the task of finding superintendents in the Australian context (Gurr, Clarke, Drysdale, & Wildy, 2014). Superintendents as they exist in contexts such as the USA, are system leaders in the sense that they exert a leadership influence across many schools. Nir (2014) indicated that superintendents can have many roles such as: acting as a mediator between competing stakeholder demands; educational, instructional, political, and managerial leaders; working in schools with principals and teachers to improve instruction, provide direction, and assist with or control human and financial resources; executive officer of a school board overseeing a school system; acting as a system leader; and so forth. In searching for superintendents, we explored the states of Victoria and Western Australia in an Australian system which comprises six states and two territories with considerable autonomy from the federal government. The choice of states was partly because they were the resident states of the authors, but also because they provided typical examples of what has been happening in Australian education. Victoria was the state with the longest history of government school self-management (see Caldwell, Hill, Gurr, & Rowe, 1997; Gurr, 1999), although from its most devolved point in 1999 when it established self-governing schools (Self-Governing Schools program expanded, News Release, Office of the Minister for Education, May 16, 1999), it had settled back to a more equal sharing of system and school responsibility for school improvement. Both states had a long history of school inspectors, with this transforming in the seventies and eighties into regional systems with regional directors, and schools gaining schools councils and increasing school autonomy (Campbell & Proctor, 2014). Similar to the Victorian initiatives of the nineties, Western Australia in recent years has become possibly the most devolved state in Australia with the establishment of Independent Public Schools in 2009. Like the Victorian self-governing schools, these schools have more flexibility in planning and day-to-day running (such as student support, staff recruitment, financial management, governance and accountability), but are still part of the government system. Since the first intake in 2010, the number of Independent Public Schools has grown to be 70 percent of all schools in Western Australia (Department of Education, 2016). In our search for superintendents, we didn't find them. Instead we argued that the superintendent role is like a chameleon in that the person and/or position that best encompasses the functions typically associated with superintendents keeps changing depending on the latest restructure and government approach to administering education.

For this chapter, we extend our ideas from Gurr, Clarke, Drysdale and Wildy (2014) and focus on research in the state of Victoria, Australia that might help in understanding the complexities of a term like system leadership. We do this from system, region and school perspectives using four pieces of related research from the University of Melbourne, and one report of district reform, to provide an integrated perspective that links system through to school level leadership. At the system level, in a doctoral dissertation, Butler (2014) explored how system initiatives and directions were filtered through layers of leadership to influence the work of school leaders. At the regional level we utilize research on a five-year improvement program in the Northern Metropolitan Region of the Victorian school system. Hopkins, Munro and Craig (2011a, 2011b) described how this region developed and applied a framework, *powerful learning*, that was adopted by most schools in the region and which led to demonstrable improvement in student learning and wellbeing outcomes. Within this region, we report on research about the construction of Hume Central Secondary College (HCSC), which was created from the closure of three failing schools. Reported in Gurr, Drysdale, Longmuir and McCrohan (2018a, 2018b), McCrohan's doctoral dissertation explored the leadership of the principal of this school though a multi-perspective case study encompassing interviews with the principal, teachers, parents, students and school council members, and observation of the school. In her master dissertation, Huerta Villalobos (2013), through interviews with the critical friends, principal and leading teachers, explored how the principal of HCSC used critical friends to help improve this school. In the final part of this chapter we relate the findings to our research on successful school leadership in schools in Melbourne, Victoria.

## System Perspectives: The Role of System Leaders and Regional Network Leaders in Improving Victorian Schools

Butler (2014) explored system leadership in the Victorian education system. At the time of the research, the Department of Education and Early Childhood Development (DEECD) had an organizational structure for schools that was headed by a Secretary, Deputy Secretary Office for Government School Education, nine Regional Directors, and then many regional network leaders (RNL), who were generally former principals that were responsible for 20–25 primary and secondary schools and their principals. Butler (2014) was interested in how system leadership was manifested and how it contributed to school improvement and he was particularly interested in the construction of the regional networks and the RNL role to supervise them (DEECD, 2008; Pike, 2008) and whether this was influencing school improvement. RNLs were a new role in the Victorian sector, and they acted in supporting principals and the school communities to improve, and also acted in a line management role for system initiatives and processes (such as supporting the school accountability process). There was, however, uncertainty about what the role could be, with conjecture and concern that it might be akin to the superintendent role in the USA, which has both trust and control elements (Nir, 2014). In a previous paper (Gurr et al., 2014) we identified RNLs as having some of, but not all, the characteristics of superintendents. Butler (2014, p. 1) described this as regulatory action "to deliberately construct a narrowly but explicitly defined system leader role within education." The study relied on system documents and individual interviews with four central senior managers (including the DEECD Secretary, and members of the senior management team of the Office of Government School Education), three regional directors, 14 RNLs and 23 principals to examine system leadership within the Victorian government education system.

The concept of system leadership explored by Butler was focused on building capacity for improvement in more than one school, and to generate change through system connections. From his research, Butler (2014, p. 96) arrived at the following definition of system leadership:

> System leadership is the ability to generate change across a system or nested system where this involves creating, utilizing or exploiting connections within the system.

It is a modest definition in many respects. Whilst it captures the core focus to improve a system of schools, it lacks mention of the typical control/power mechanisms such as governance, fiscal and human resource management, direction setting, accountability and so forth. It is, however, an inclusive definition in that it allows many people at different levels of a school system to demonstrate system leadership. In Butler's research, senior managers were clearly viewed as system leaders. There were also expectations and indications that RNLs could be system leaders, but many interviewees also viewed the RNL role as being more concerned with line management and compliance. The research did indicate the potential of this role to enhance horizontal/heterarchical leadership. Principals were not generally seen to be system leaders, primarily because of their single school focus, and also because of competitive pressures not to support the work of other schools. The evidence that system leadership led to school improvement relied more on argued cases of indirect impact, rather than clear empirical evidence.

In the current Victorian system, system leadership according to Butler's (2014) definition would still most easily be found within the senior management roles, especially that of Secretary of the Department of Education and Training (DET), the Deputy Secretary Regional Services Group, and the four regional directors. They all have budgetary controls, and are involved in setting system directions, monitoring accountability, supporting building improvement and so forth. They are also heavily involved in policy and process developments that lead to change "across a system or nested system where this involves creating, utilizing or exploiting connections within the system." It should be noted that in Victoria, school employment has long been devolved to the school level, including the appointment of a principal. These senior managers are all somewhat removed from schools, and none would work directly with schools on a regular basis. The work undertaken by Senior Education Improvement Leaders (SEILs), which are the current version of RNLs, typically engages them with schools on an almost daily basis, but is generally focused on working with principals and other senior school leaders. Many are also involved in promoting change amongst their network of schools, and all are involved in principal performance appraisal and school accountability. However, as with RNLs, SEILs are not involved in setting system directions, and they have no direct fiscal, building or human resource control (other than principal appraisal).

In interpreting Butler's (2014) research, we described how the work of RNLs relied more on the trust dimension of superintendent work described by Nir (2014), as they did not have sufficient power to engage much with the control aspect (see Gurr et al., 2014). Conversely, senior department managers relied more on control than trust, as they are too removed from schools to establish the type of relationships needed to promote trust. The exception is that of the regional director role. Some of these directors have in the past established strong, positive relations with schools and so can use both the trust and control dimensions of the superintendent role described by Nir (2014), as they have sufficient influence in aspects such as direction, budgets, buildings, and accountability to exert control over schools, whilst they are close enough to schools to establish positive relationships that lead to using trust for influencing schools. We now explore this role further when we consider the work of Wayne Craig in the Melbourne Northern Metropolitan Region.

## Regional Perspectives: The Improvement of the Northern Region of Melbourne

The edited book by Hopkins, Munro and Craig (2011a) described the school improvement journey of the former Northern Metropolitan Region led by

Wayne Craig as the Regional Director. The book tells the story of how Craig led the improvement of this region through the development of a school improvement framework, *Powerful Learning*, the use of experts in literacy, numeracy, student welfare, and system leadership, and the support of RNLs, principals, and teachers in a collective effort to improve student learning.

At the time of the reform the Northern Region included 195 schools comprising 137 primary schools, 36 secondary schools, 13 special schools, seven primary and secondary schools, one school that went from primary years to year 9 and one school that only had years 10–12. These schools served 75,000 students and some of the areas had some of the highest levels of social disadvantage in Australia.

Craig inherited a region that was one of the lowest performing in the state and set about, with the help of Hopkins (school improvement) and Munro (pedagogy), to construct an improvement process centered on what they termed Action Improvement Zones – AIZ (Hopkins, Munro, & Craig, 2011b). They enlisted support from Lewis (2011) for student welfare, and Sullivan (2011) for mathematics. The booklet *Powerful Learning* (NMR, 2009) summarized the approach to the AIZ through a succession of circles of practice beginning at the center with the intention to develop students that were literate, numerous and curious. The model is shown in Figure 12.1.

This inner circle aligned with national statements about schooling, such as the Melbourne Declaration (Ministerial Council on Education, Employment,

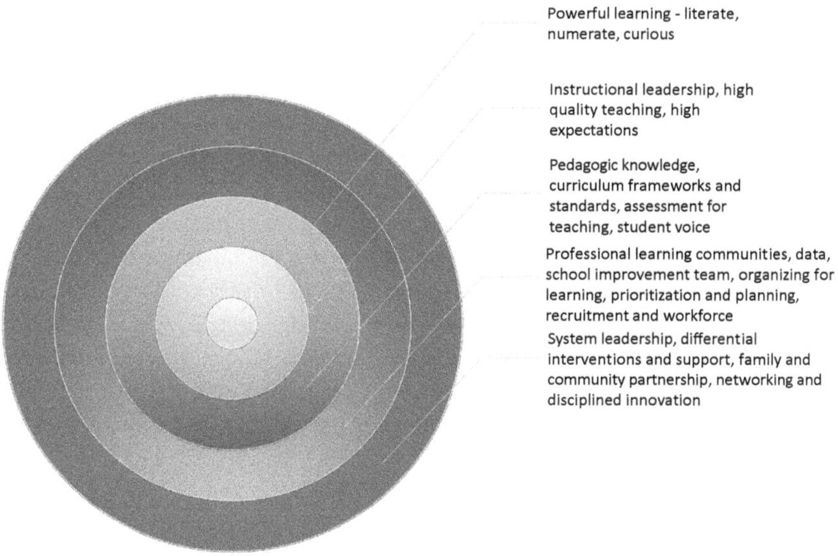


**FIGURE 12.1** Northern Metropolitan Region School Improvement Model (adapted from NMR, 2009, p. 11)

Training & Youth Affairs, 2008). The next circle included three pre-conditions for school effectiveness; the high leverage learning improvement strategies of instructional leadership, high quality teaching and high expectation. The focus then was on the classroom by developing quality teaching and learning through pedagogic knowledge, curriculum frameworks and standards, assessment of teaching and student voice. Schools supports (organizational capacity) for improving teaching and learning then followed by focusing on professional learning communities, collection and use of data, school improvement teams, organizing for learning, prioritization and planning, and recruitment and workforce planning. The outer circle focused on the systemic context through considering big picture and external supports for schools such as system leadership, differential school improvement intervention and support, family and community partnership, and networking with other schools and disciplined, evidence-informed innovation. For system leadership, whilst there was no formal definition given, the system leadership initiatives described in this outer circle focused on getting principals to influence the improvement of many schools, and to support this work through system leaders (like RNLs).

Each circle included a description of the key ideas, expected practices, and statements of intent as to what the region would provide. This strategy connected with already developed system initiatives on effective schools, student engagement (E5 model), principles of teaching and learning, effective professional learning, and school leadership.

The book included chapters from all the key players – Craig, Hopkins, Lewis, Munro and Sullivan – as well as contributions from Fraser, who was the Deputy Secretary Office for Government School Education, and a highly regarded educational consultant, Zbar (who was also one of the critical friends to Hume College described below). These chapters provided the details as to what happened. From a system leadership perspective, it is important to recognize that Craig, as the regional director, had considerable control over schools, and that through the construction and articulation of this initiative, he gained trust. One of the authors of this chapter, Gurr, was conducting school reviews in this region during this reform, and can attest to how schools used the language and resources of the AIZ/Powerful Learning program, the high regard they had for the clarity this initiative brought to the school improvement process, and the active role of RNLs to support schools to implement the reform. Importantly, the book included a chapter that presented some evidence of positive change trends (Fraser, Glover, & Craig, 2011). Glover led, and Fraser had a senior leadership role, in the DEECD Data, Outcomes and Evaluation section, and were responsible for the school review process. The chapter used a range of school data collated at the system level (including student learning outcomes in literacy and numeracy (standardized tests and teacher judgements), survey data from students, parents and teachers, and student pathway and transition data (retention, and student destinations on leaving school). They concluded (Fraser et al., 2011, p. 151) that:

> The overall conclusion to be drawn from the review of data in this chapter is that over the past four years there has been a quite dramatic shift in the metrics from a largely negative to a strikingly positive direction. In particular, literacy and numeracy measures for Years 3, 5, 7 and 9 are at or near state benchmarks with the data generally trending upwards.

Whilst the performance in literacy and numeracy were particularly pleasing, some of the other data sets were showing positive trends but without substantial gains by the time the chapter was written. So, in some ways it is a generous conclusion. Nevertheless, they did make an argued case that the NMR improvement strategy compared favorably with best practice initiatives worldwide, noting in particular that there was (Fraser et al., 2011, p. 152):

- A clear and comprehensive model of reform
- Strong leadership at the regional level
- Substantive training related to the goals of the program
- Implementation support at the school level
- An increasingly differentiated approach to school improvement

This suggested that the reforms would continue to show improved school outcomes. Unfortunately, the reform continued for only a few more years without any major evaluations, and so it is difficult to judge its sustained success.

## School Perspectives: The Improvement of HCSC

This section explores the improvement journey of HCSC and focuses on the school's leadership in a devolved system. We trace the establishment of a new school born from closure of three failing schools in a disadvantaged area, to try to explain its improvement over six years. In analyzing and exploring the school's improvement from the school's perspective we use a framework which is adapted from Hallinger's (2016) instructional management model. We draw on research from Gurr, Clarke, Drysdale and Wildy (2014) involving interviews with the principal, Glenn Proctor, and Gurr, Drysdale, Longmuir and McCrohan (2018a, 2018b), which employed a multiple-perspective qualitative methodology that included interviews with the principal, campus principals, members of the leadership team, middle-level leaders, teachers, students, parents, and school council members concerning the improvement of the school; this research is part of the International Successful School Principalship Project (ISSPP) that we have been part of since its beginning in 2001. Also, one of the authors, Drysdale, worked at the school as a critical friend from the beginning of the new school (again, the role of the critical friend at this school has been explored by Huerta Villalobos, 2013, and is discussed below).

## *The* School Leadership in Context *Framework*

Models or frameworks help to organize key variables and concepts in an attempt to map the relationships between these variables. In the *School Leadership in Context* framework (Figure 12.2), we build on the work of Hallinger (2016) to explain school improvement in an educational system. We have made some significant modifications to the model in an attempt to take a school perspective on system leadership.

In Figure 12.2, we identify four general contextual factors that impact on the school: economic, socio-cultural, political, and technological. We also recognize that the school exists within a broader educational system where policy, system initiatives, regulations, and guidelines can influence the school (as outlined in the previous sections). Hallinger (2016) classifies this as the institutional context. Within the educational system or institutional context is the school perspective. In the center of the school perspective is the school performance and development context with leadership as the central feature. Leadership influences school culture and climate, and teaching and learning, which in turn impacts on student outcomes. Our model identifies three direct contextual influences (school characteristics, personal leadership characteristics, and community context) that influence the performance and development context and leadership. In addition, we include support from external agencies and networks which influence leadership directly and school improvement indirectly.

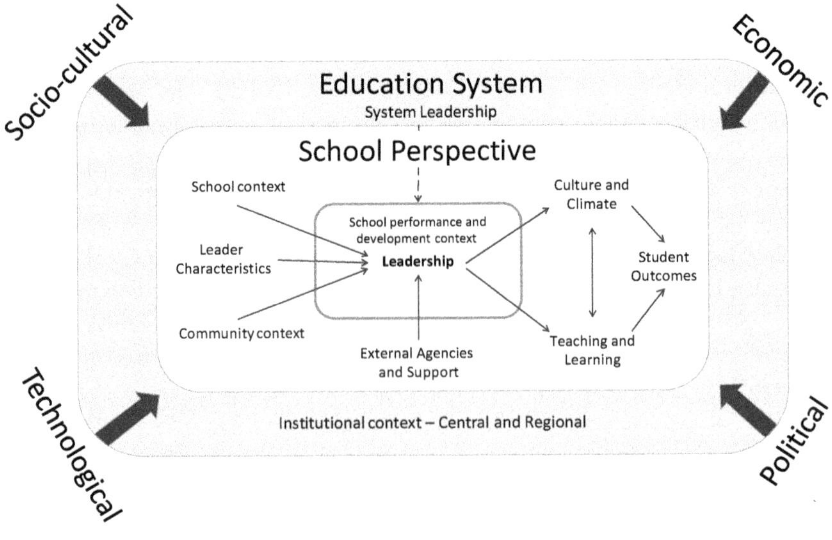


**FIGURE 12.2** School Leadership in Context (adapted from Hallinger, 2016, p. 13)

In what follows, we trace the school's improvement journey using the model. We explore the school's perspective in the following order: system or institutional context; the three direct contextual influences (school context, leader characteristics, community context); external networks and agencies; and the school's performance and development context and school leadership. We acknowledge the importance of the general environment factors (socio-cultural, economic, political and technological), but do not discuss these, and instead limit ourselves to those contexts that directly influenced the school's improvement trajectory.

### *System or Institutional Context*

The institutional context incorporates the location of a school within a system or systems of education. As noted previously, the Victorian government system was an early adopter of school self-management, and was characterized by a high level of school autonomy and flexibility. In the case of HCSC, this allowed the school principal to undertake significant change at the college. In establishing a new school from the ashes of three failing schools, Glenn's mandate was to set about establishing a new school philosophy, policies, processes, programs and practices.

Having said that, Glenn sought guidance from central and regional personnel. In particular, Glenn relied on the Regional Director, Wayne Craig, to support decisions that were integral to the successful foundation of the school, such as extending the contract of key staff that were on loan from other schools, and introducing a minimum attendance expectation on students to make them more personally accountable for their learning. Glenn had extensive experience working as a school principal, and his familiarity with the system was essential in understanding when he required the regional director's support, and when he could utilize the autonomy available within a Victorian government school. From a system leadership perspective, the role of regional personnel was more to support Proctor's work than to intervene or control what was happening. So, it was a light-touch form of system leadership, and highlights how successful principals often do not need close supervision from systems.

### *School Context*

Glenn Proctor was appointed as the executive principal of HCSC in 2008, although the school was not officially opened until 2009. The establishment of the new school was part of a government regeneration project in the Northern Region (described in the previous section) that was aimed at transforming the educational opportunities and achievement levels for students in one of the most disadvantaged communities in Australia with 75 percent of students from the bottom quartile of socio-economic advantage (Department of Health and Human Services, 2014).

Glenn, as executive principal, was responsible for closing the campuses of three separate 7–12 colleges, and establishing the new school. The three schools had suffered from decades of neglect, poor leadership, declining enrolments, low student performance (some of the worst in the state), negative school cultures, low staff morale, and disruptive and disconnected students. The schools' reputations were such that most families in the area passed by the schools to go somewhere else. Glenn had to work with the existing staff and leaders from the three schools, and found himself responsible for the construction of three new campuses, with two situated on new sites. The new school was a Year 7–12 co-educational secondary school consisting of two Year 7–9 campuses and one Year 10–12 campus. The new buildings, completed in 2011, were in striking contrast to the previous school buildings which were tired and poorly maintained.

In 2009 there were 1,000 students and enrolments were projected to decline. Staffing included an executive principal, three campus principals and 108 teachers. By 2015 the enrolment was 1,125 (508 girls and 617 boys) with projection for further growth, and there was an executive principal, six other principal class personnel, and 131 teaching staff.

### *Personal Leadership Qualities*

Glenn had more than 35 years' experience in schools, beginning as an economics and accounting teacher before moving into school leadership. Prior to arriving at HCSC he was principal of Mount Waverley Secondary College for 11 years, which is a high performing school in an affluent suburb of Melbourne. However, Glenn grew up in Broadmeadows, and so he had an affinity with the area and an intimate understanding of the context of HCSC.

Glenn demonstrated the characteristics of what we have termed the post-heroic leader (Drysdale, Bennett, Murakami, Johansson, & Gurr, 2014). Initially he began as a directive leader and showed many of the characteristics of typical heroic leadership (Adair 1989, p. 227). He set clear improvement directions, challenged the status quo, and showed courage to stand up to any in power and authority who may have questioned the direction he was taking. Throughout his principalship he had a clear social justice focus and demonstrated a strong commitment to improve the education of students in challenging circumstances. His general and educational values never wavered. He showed respect for others, strongly advanced social justice and equity issues, took personal responsibility for his actions, and held firm to the beliefs that every student could learn and wanted the opportunity to succeed. He showed integrity by modeling the way forward, and putting duty before self. After gaining some initial success (such as improving student attendance which increased from a low of 60 percent in 2009, to eventually reach 89 percent in Glenn's final year, 2016), he changed to a more collaborative and post-heroic style of leadership that involved the leadership of many. An example is how he recognized early that there was a need to build the leadership capacity of his principal leadership team. He did this through coaching (the *Coaching for Success* program), as well as targeted professional learning opportunities, and using research that supported the school's context and improvement trajectory. Glenn was able to adapt his leadership to the circumstances, and more recently we have described Glenn as a transformational and somewhat disruptive leader (Drysdale, Gurr, & Goode, 2017; Drysdale, Gurr, & Longmuir, 2017), who was good at motivating, understanding and developing staff, and who was also looking to promote innovation and change.

### Community Context

Community context considers the school within the community it serves. The community of HCSC was made up of students mainly from the suburb in which the school was located. The initial observations of HCSC made by Glenn following his appointment was that there were many students walking past the school to go to other schools. The school enrolment numbers were low, and students experienced disruption to their learning due to misbehavior and poor attendance. Except for those parents who sat on the school council, there had been a history of little, or no, parental involvement in the three schools prior to amalgamation. The low level of parental engagement did little to support the vision of the college. As a result, Glenn identified the need for greater connection between the school and home to promote aspects that would help students succeed, such as coming to school; as already mentioned, some students' attendance was as low as 60 percent, and one of the amalgamated schools had an average student absence rate of 35 days. Greater accountability for students to attend was implemented (there were minimum attendance requirements to ensure promotion to the next year level), along with clear structures for teachers to follow in the event of low attendance. Students were also expected to arrive at school on time and be punctual to classes. A system called "Time Counts" was developed. Time Counts required students to be seated in class by 8:45am. If a student was marked as late three or more times in a week, they were required to work for 30 minutes of their lunchtime on missed learning. During an observation of the school, staff members on duty at the beginning of the school day regularly announced, "time counts", as a signal to the students to move to class quickly. This, along with the expectation of regular attendance at school, quickly changed students' perception of the importance of their education and they became more accountable for their learning.

### School Performance and Improvement Context

This is related to the historical context of the school and the school's improvement trajectory. Hallinger (2016) notes that this context also defines the nature of the principal's challenge. Given that HCSC was the result of the closure of three failing schools, it was not surprising to say that the school had a history of poor performance, inadequate facilities and decades of neglect. Gurr reviewed one of the three closed schools in the early 2000s and judged the school to be one of the lowest performing schools in the state. When HCSC began there were extremely low academic standards, poor attendance, few students going on to tertiary education, and limited aspirations among students: "They were at the bottom of the tables academically; nearly half the students weren't even turning up, and only 30 per cent finished year 12" (News Documentary, *Four Corners ABC*, 2011).

One of Glenn's first priorities was to set about establishing a new history, direction and culture. His strategy for improvement was to engage students, build staff capacity, improve teaching and learning practice, raise staff and student expectations, develop a new positive school culture, more effectively manage resources, and introduce a higher level of staff accountability for student learning. To engage students, he created Curriculum Design Teams (CDT) to develop a guarantee and viable curriculum, and promote appropriate assessment practices, such as moderation between teachers. To set high expectations he developed a 2:1 strategy in which the school tried to provide two years of learning growth for each calendar year. To build leadership capacity he invested significant resources in developing a high performing leadership team that included senior and middle-level leaders (see below). To support the work of teachers to improve teaching and learning, Glenn focused on developing professional practice, in general, and purposeful teaching, in particular. A key strategy was initiating a common instructional model, and establishing peer coaching and classroom observation to build collaborative practices and a culture of relentless improvement.

### External Agencies and Networks: The Role of Critical Friends

External agencies and networks played an important role in helping to influence the school performance and improvement context. The school acknowledged the support of agencies such as the charity organization, The Smith Family, which worked in partnership with the college to facilitate programs to support student learning, especially for those students who had difficult family circumstances. The school worked with the Technical and Further Education (TAFE) sector to provide vocational programs and facilities. The school partnered with several primary schools to ensure transition to secondary school was smooth and effective. Local business also supported the college with on-the-job training and work experience for students.

While these initiatives were important, the use of several critical friends was crucial to the school's improvement. Critical friends can provide professional support, advice, reflection, but also question and challenge assumptions and practices. It is not a formal role, such as a mentor or coach, but more a professional relationship based on mutual regard, respect and trust. They can offer a critical perspective and another lens on the school. Huerta Villalobos (2013) explored the role critical friends played in the school's improvement.

> The role of the critical friend was found to be a dynamic one, requiring a high level of skill, flexibility, and professional judgement. Rather than following a checklist of scripted "technical assistance", it was about developing a repertoire of strategies and skills, and learning when and how to use them, taking account of the context.
>
> *(Huerta Villalobos, 2013, p. 68)*

There were two critical friends that greatly impacted on the school leadership team and the principal. Educational consultant Vic Zbar was engaged to work with the leadership team to implement a framework of school improvement based on his research on successful schools in educationally disadvantaged areas. Vic was employed on a regular basis to support the improvement agenda. Lawrie Drysdale, a co-author of this chapter, was engaged (voluntary) as a critical friend from 2009 to 2015. He regularly attended senior leadership meetings and conducted a program for emerging leaders for five years. His insight into the role was captured by Huerta Villalobos (2013) who conducted a single-site multiple perspective case study involving interviews with 13 people: two critical friends, the executive principal, three campus principals, one assistant principal, six teacher members of the leadership team. Questions centered on participants' perceptions of the role and impact of the critical friends. She found the critical friends had a direct impact on the work of senior and middle level leaders, and through this, an indirect impact on the work of teachers and student outcomes. They were considered by participants to be more important and influential in school improvement than would be the case if they were an internal coach or external agent working with the school. Their trusted and acknowledged expertise, combined with their close connection with the school, promoted a more influential role within the school. This study of the work of critical friends highlighted again the potential for extensive control of improvement at the school level, and the lesser role of system players. This is further highlighted by the following reflections of Drysdale.

## *Reflection of a Critical Friend: System Leadership and the Principal*

As a critical friend Drysdale was privileged with insight and information that provided a unique view of the principal's approach to school improvement. Drysdale was a participant observer who kept records of meetings and programs he attended with the leadership team, and reflections on personal conversations with the principal. There was a professional relationship with the principal that spanned over 10 years (five years at Mt Waverley SC and five years at Hume CSC). As such there was a high level of trust between the principal and Drysdale.

Glenn fully understood that he operated within a system. There were many actors that were in the system that the principal and school could draw upon, mainly at the regional level; the reader will recall that the Northern Region was engaged in a major school improvement effort (refer to Figure 12.1). Glenn acknowledged system initiatives and policy changes, but set his own course in establishing the school and fulfilling his mission to establish a successful school. Glenn used the resources provided by the system where he saw these as advantageous. The system had invested tens of millions of dollars in the regeneration project and was determined that the project should succeed. Glenn was given a great deal of autonomy to achieve this; however, he did use his personal relationship with the regional director, Wayne Craig, to extract additional resources, and authority around working with existing staff and recruiting new staff. He kept the regional director informed, and invited him to key school events. Glenn also engaged in regional initiatives and programs. However, he was circumspect with regard to involvement with other regional personnel, and had a healthy skepticism of system support. He only engaged with those he trusted and whom he believed could add value to the school. Glenn's general approach was to focus on the school. He often refused to attend meetings at regional and system level if they were to take him out of the school too often. He believed that if he was to be an effective leader he needed to be visible in the school. He had an office in each campus where he spent time each week. Leadership meetings were rotated between campuses. As a visible leader, system leadership was only important in so far as it supported the school's improvement agenda. His improvement focus was strongly augmented by external agencies and networks, such as the Hume Council.

More than any type of system leadership, Glenn placed his faith in evidence-based decision making. Glenn was comfortable to declare that he had "never had an original idea in his life." This humble declaration reflected his commitment to evidence-based decision making. He engaged with renowned and prominent academics and researchers. This was particularly evident through his decision to focus on areas that research identified as key to supporting school improvement, such as strong leadership that is shared, high expectations, teacher efficacy, ensuring an orderly school environment, and focusing on key priorities that were drawn from research evidence.

Drysdale was consulted about the agenda for senior leadership meetings. Team meetings would run for 90 minutes and focused on professional development. Typically, meetings included one or more of the following activities: workshops, skills training, discussion of a reading, presentation by members of the leadership team on a relevant topic, analysis of student or school data, presentation of material from a respected expert or academic. Drysdale was regularly requested to contribute articles and materials to support a theme or topic that Glenn had decided would build leadership capacity, such as leading effective teams, instructional leadership, and managing conflict. Drysdale's role varied in the meetings he attended: he might comment on a staff presentation or strategy; run workshops of important topics such as middle-level leadership; lead discussion on a reading; participate in skills training like active listening, give feedback to staff who led discussion on a topic; or, provide feedback to the principal after the meeting. Occasionally regional officers and senior officers were invited to attend the sessions to inform them of the school's progress.

Glenn introduced two key leadership programs, Coaching for Success and Coaching for Performance, run by the local TAFE. These two major initiatives formed the backbone of the senior leadership's training for five years, and Drysdale participated in them. Skills practice related to these programs was often included as a component of the leadership meetings. Drysdale also conducted an emerging leadership program aimed at building the capability of the next generation of leaders. This program resulted in several high potential leaders being promoted to leadership positions.

## Reflection

This paper has described several empirical studies within the one jurisdiction, with these studies going from system to school level perspectives. Using Butler's (2014) definition of system leadership, at the system level we saw that there were people that seemed to operate as system leaders, and these were at senior levels of the central and regional systems, with RNLs having the potential to exercise system level leadership. Principals tended to not operate as system leaders because they had limited influence across schools. At a regional level, it was clear that regional directors could act as system level leaders, exerting wide influence on clusters of schools to improve. When we moved to the school level we saw in the case of one successful principal that system leadership was helpful, but not the most important influence on school success. Whilst it could be helpful and act as a support, the work of the principal, other school leaders, and, in the case study school, critical friends, were more important to the improvement journey of the school. So, what the reader perhaps can see is that system leadership can be important, but that it needs to work in conjunction with school leadership. Without good school leadership, the level of impact of system leadership is likely to be limited.

In an OECD review of school leadership practices involving 22 country reports and five detailed country case studies, Pont, Nusche and Hopkins (2008, p. 9) claimed that:

> One of school leaders' new roles is increasingly to work with other schools and other school leaders, collaborating and developing relationships of interdependence and trust. System leaders, as they are being called, care

> about and work for the success of other schools as well as their own. Crucially they are willing to shoulder system leadership roles because they believe that in order to change the larger system you have to engage with it in a meaningful way.

They cited Belgium, England, and Finland as examples of systems that have encouraged principals to cooperate with other principals. In our past research in the Victorian context we can see little evidence of the presence of this type of leadership operating at the principal/school level. Despite various system initiatives over time to promote greater cooperation between principals in Victorian government schools, the competition between schools for enrolments (Bentley & Butler, 2017) and intense workload and high stress (Riley, 2017) work against these efforts. However, our two most recent cases in the ISSPP (Longmuir, 2017) are both government school principals and provide some evidence of system leadership. Both have led the significant improvement of schools in affluent areas that were not performing at expected levels. One of the principals, Robyn (not her real name), has exerted a minor system leadership role in terms of participation in system-wide committees whilst she was a principal; prior to working at this school, she had been a system level curriculum specialist and so her committee work was in some ways an extension of her previous work across schools. However, her main contribution at system leadership level was when she took a short-term secondment to work as a SEIL for several months (the reader will recall that RNL/SEILs were shown by Butler (2014) to have a system leadership role). For this principal, for her to have a significant system leadership role meant stepping outside of the principalship. The other principal, Michael (not his real name), led what has become acknowledged as a very innovative and strongly student-focused school. The school had gained such a reputation that Michael formed user-pay professional learning days to allow people to explore what the school was doing. He had influenced at least two other schools that we know of and had influenced these schools through principal collegial support and professional learning opportunities for staff of these schools. This support allowed these schools to model their schools' improvement on Michael's schools. It is important to note, however, that neither of the schools were near Michael's school, and so they were not competitor schools. Michael has now resigned his principalship and become a private consultant in the belief that this will allow him to best promote his vision for schools. Whilst Michael, in particular, is influencing other schools, if we consider again Glenn Proctor, he was very clear that his focus had to be on his own school and not that of helping other schools. He stressed the importance of principal visibility in the school, and was totally focused on his school's improvement. Whilst he was well connected with the region in particular, and used his networks for the advantage of his school, he was not interested in influencing beyond his school. His example is more typical of what we observe in our work with principals in Victoria. Principals will participate in external system committees when asked, and some will voluntarily be involved in professional associations and so forth, but few are extensively involved in supporting the development of other schools whilst they are working as principals.

When we consider our own extensive research on school leadership in Victorian schools through projects like the ISSPP, there has been very little consideration of the positive impacts of systems. Indeed, in our successful school leadership research, our principals have often described how they had to shape system behavior so it didn't impact negatively on their school. An example is the leadership of Jan Shrimpton (Drysdale, Goode, & Gurr, 2009, 2011; Goode, 2017). Jan had turned around two failing schools in her career. At her last school, she had got the school to a level of performance that was above expected levels. There was a push from the system for her to get an even higher level of student learning performance. She resisted this, because for her to get further improvement in literacy and numeracy meant compromising other areas of the school's success. For example, further improvement in literacy might have required more time, compromising time for other curriculum areas. In the leadership models we have produced (Drysdale & Gurr, 2011; Gurr, 2015), and in our discussion of context and leadership (Gurr, 2014; Gurr et al., 2018a, 2018b), we described how our successful leaders worked with and changed context to benefit their schools. So, it is not that they will be against the system leadership efforts of senior bureaucrats, but that they will modify/adapt the mandates to suit their school needs, and may even try to influence the system to provide a climate more suited to what they need to promote school success. This is activist and somewhat heroic leadership (Drysdale et al., 2014).

## References


Adair, J. E. (1989). *Great Leaders*. Guildford, UK: Talbot Adair.

Bentley, T., & Butler, S. (2017). Collaboration in pursuit of learning. In T. Bentley & G. C. Savage (Eds.), *Educating Australia: Challenges for the decade ahead* (pp. 296–312). Carlton, Australia: Melbourne University Press.

Butler, S. (2014). *System leadership in Victoria, Australia*. Unpublished doctoral dissertation, The University of Melbourne.

Caldwell, B. J., Hill, P. W., Gurr, D., & Rowe, K. J. (1997, March). *The Schools of the Future program in Victoria, Australia: The principal's perspective*. Paper presented at the American Educational Research Association National Conference, Chicago, IL.

Campbell, C., & Proctor, H. (2014). *A history of Australian schooling*. Sydney, Australia: Allen and Unwin.

Department of Education (2016). *Opportunity to become an independent public school in 2017*. Perth, WA: Western Australian Department of Education.

Department of Education and Early Childhood Development (DEECD) (2008). *Blueprint for Education and Early Childhood Development*. Victoria, Australia: DEECD.

Department of Health and Human Services (2014). 2014 town and community profile for Broadmeadows (suburb). Retrieved from: https://www.data.vic.gov.au/data/dataset/2014-town-and-community-profile-for-broadmeadows-suburb

Drysdale, L., Bennett J., Murakami, E., Johansson, O., & Gurr, D. (2014). Heroic leadership in Australia, Sweden, and the United States. *International Journal of Educational Management, 28*(7), 785–797.

Drysdale, L., Goode, H., & Gurr, D. (2009). An Australian model of successful school leadership: Moving from success to sustainability. *Journal of Educational Administration, 47*(6), 697–708.

Drysdale, L., Goode, H., & Gurr, D. (2011). *Sustaining school and leadership success in two Australian schools.* In L. Moos, O., Johansson, & C. Day (Eds.), *How school principals sustain success over time: International perspectives* (pp. 25–38). Dordrecht, The Netherlands: Springer-Kluwer.

Drysdale, L., & Gurr, D. (2011). The theory and practice of successful school leadership in Australia. *School Leadership and Management, 31*(4), 355–368.

Drysdale, L., Gurr, D., & Goode, H. (2017, November). *Leading the rebuilding of schools through disruptive innovation and leadership.* Paper presented at the University Council for Educational Administration (UCEA) conference, Denver, CO.

Drysdale, L., Gurr, D., & Longmuir, F. (2017, November). *Innovative leadership in underperforming schools.* Paper presented at the Australian Association for Research in Education (AARE) Conference, Canberra, Australia.

Fraser, S., Glover S., & Craig, W. (2011). The NMR performance story within the broader context of state-wide data development and school improvement policy. In D. Hopkins, J. Munro, & W. Craig (Eds.), *Powerful learning: A strategy for systemic educational improvement* (pp. 127–152). Melbourne, Australia: ACER.

Goode, H. (2017). *A study of successful principal leadership: Moving from success to sustainability.* Unpublished doctoral dissertation, The University of Melbourne.

Gurr, D. (1999). *Reforms in school supervision and quality control: The case of the state of Victoria.* Paris, France: International Institute for Educational Planning.

Gurr, D. (2014). Successful school leadership across contexts and cultures. *Leading and Managing, 20*(2), 75–88.

Gurr, D. (2015). A model of successful school leadership from the International Successful School Principalship Project. *Societies, 5*(1), 136–150.

Gurr, D., Clarke, S., Drysdale, L., & Wildy, H. (2014). Looking for Superintendents in Australia, the cases of Victoria and Western Australia. In A. E. Nir (Ed.), *The educational superintendent between trust and regulation: An international perspective* (pp. 113–130). Hauppauge, NY: Nova Science Publishers.

Gurr, D., Drysdale, L., Longmuir, F., & McCrohan, K. (2018a – in press). Leading the improvement of schools in challenging circumstances. *International Studies in Educational Administration, 46*(1).

Gurr, D., Drysdale, L., Longmuir, F., & McCrohan, K. (2018b – in press). Successful school leadership that is culturally sensitive but not context constrained. In E. Murakami, D. Gurr, & R. Notman (Eds.), *Leadership, culture and school success in high-need schools.* Washington, DC: Information Age Publishing.

Hallinger, P. (2016). Bringing context out of the shadow. *Educational Management Administration & Leadership*, pp. 1–20, article first published online: December 2, 2016.

Hopkins, D., Munro, J., & Craig, W. (Eds.) (2011a). *Powerful learning: A strategy for systemic educational improvement.* Melbourne, Australia: ACER.

Hopkins, D., Munro, J., & Craig, W. (2011b). Introduction. In D. Hopkins, J. Munro, & W. Craig (Eds.), *Powerful learning: A strategy for systemic educational improvement* (pp. 1–7). Melbourne, Australia: ACER.

Huerta Villalobos, M. (2013). *The role of the critical friend in leadership and school improvement*. Unpublished Master thesis, The University of Melbourne.
Lewis, R. (2011).The Developmental Management Approach to classroom behaviour. In D. Hopkins, J. Munro, & W. Craig (Eds.), *Powerful learning: A strategy for systemic educational improvement* (pp. 96–108) Melbourne, Australia: ACER.
Longmuir, F. (2017). *Principal leadership in high-advantage, improving Victorian secondary schools*. Unpublished doctoral dissertation, The University of Melbourne.
Ministerial Council on Education, Employment, Training & Youth Affairs (2008). *Melbourne declaration on educational goals for young Australians*. Melbourne, Australia: Ministerial Council on Education, Employment, Training & Youth Affairs.
Nir, A. E. (2014). Building trust and control: A challenge for school superintendents. In A. E. Nir (Ed.), *The educational superintendent between trust and regulation: An international perspective* (pp. 1–16). Hauppauge, NY: Nova Science Publishers.
Northern Metropolitan Region (NMR) (2009). *Powerful learning*. Melbourne, Australia: Department of Education and Early Childhood Development.
Pike, B. (2008). *New leaders for local school networks*. Media release, 11 September 2008, from http://www.premier.vic.gov.au/minister-for-education/new-leaders-for-local-school-networks.htm (no longer available).
Pont, B., Nusche, D., & Hopkins, D. (2008). Improving school leadership, Volume 2: *Case studies on system leadership*. Paris, France: OECD.
Riley, P. (2017). *Australian principal occupational health, safety and wellbeing survey: Executive summary, 2016 Data*. Victoria, Australia: Australian Catholic University.
Sullivan, P. (2011).Ways of working with numeracy and mathematics leaders to improve student learning. In D. Hopkins, J. Munro, & W. Craig (Eds.), *Powerful learning: A strategy for systemic educational improvement* (pp. 82–95). Melbourne, Australia: ACER.

# INDEX

For Product Safety Concerns and Information please contact our EU
representative GPSR@taylorandfrancis.com
Taylor & Francis Verlag GmbH, Kaufingerstraße 24, 80331 München, Germany

www.ingramcontent.com/pod-product-compliance
Ingram Content Group UK Ltd.
Pitfield, Milton Keynes, MK11 3LW, UK
UKHW022029061126
477531UK00003B/110
*9781138559912*